普通高等教育"十三五"规划教材

电气信息类科技英语教程
第 2 版

主 编 何 宏

机械工业出版社

本书是为高等院校电气信息类专业学生编写的专业英语教材。全书采用了以科技语言翻译、写作技巧为主要学习内容，电气信息类专业阅读材料为辅助内容的方法编写。主要包括科技英语英译汉翻译技巧、写作和科技英语汉译英翻译技巧三部分。其中英译汉部分为基础部分，共12个单元，除了翻译方法外，还含有24篇科技阅读材料，内容涉及电子、通信、自动化、电气等多个领域，并包括人工智能、物联网、工业4.0、大数据、机器人等最近几年信息领域的技术热点。最后8单元汉译英和专业词汇为扩展部分，使用者可以根据需求灵活选择。

本书的内容既有针对性，又适用范围较广，可选择性强，可以作为电子工程、通信工程、自动化和电气工程专业的本科或研究生专业英语课程教材，也可供相关科技人员学习和参考。

图书在版编目（CIP）数据

电气信息类科技英语教程/何宏主编．—2版．—北京：机械工业出版社，2018.3

普通高等教育"十三五"规划教材

ISBN 978-7-111-59352-2

Ⅰ.①电… Ⅱ.①何… Ⅲ.①电气工业 – 英语 – 高等学校 – 教材②信息技术 – 英语 – 高等学校 – 教材　Ⅳ.①TM②G202

中国版本图书馆CIP数据核字（2018）第044881号

机械工业出版社（北京市百万庄大街22号　邮政编码100037）
策划编辑：王雅新　　责任编辑：王雅新　杨　洋
责任校对：王雅新　　封面设计：张　静
责任印制：张　博
三河市宏达印刷有限公司印刷
2018年4月第2版第1次印刷
184mm×260mm·14印张·337千字
标准书号：ISBN 978-7-111-59352-2
定价：34.00元

凡购本书，如有缺页、倒页、脱页，由本社发行部调换

电话服务　　　　　　　　　　　　网络服务
服务咨询热线：010-88379833　　机 工 官 网：www.cmpbook.com
读者购书热线：010-88379649　　机 工 官 博：weibo.com/cmp1952
　　　　　　　　　　　　　　　　教育服务网：www.cmpedu.com
封面无防伪标均为盗版　　　　金　书　网：www.golden-book.com

第 2 版前言

本书是在《电气信息类科技英语教程》第 1 版的基础上修改而成的。为了适应社会的需求，激发学生学习科技英文的兴趣，使学生的阅读翻译水平跟随科技发展而提高，第 2 版教材更新了 PART II 写作部分和 PART I 英译汉每一单元 C 部分的英文翻译阅读文章。鉴于目前我国大学工科学生普遍汉译英能力较差，还扩充了三个单元的汉译英的内容，以便加强学生汉译英能力的培养。此外，因为不少大学已经开设英语口语课程，所以本书第 2 版删除了第 1 版中 PART I 每个单元的 D 语音部分内容。

第 2 版教材更加强调英译汉和汉译英能力的培养。PART I 英译汉部分压缩为 12 个单元，保留第 1 版教材的英译汉理论部分，对部分单元 B 部分的专业领域英文翻译阅读进行了更新，而且还更换了每一单元 C 部分的英文翻译阅读，所选文章内容为人工智能、物联网、工业 4.0、大数据、机器人这些最近几年信息领域的技术热点。PART II 实用的写作技巧部分更新了摘要写作部分。PART III 汉译英部分补充了 Unit 6 定语从句的翻译（Translation of Attributive Clauses）、Unit 7 状语从句的翻译（Translation of Adverbial Clauses）和 Unit 8 句子中的省略（Omission in Sentence Translation）。精简了附录中的专业词汇，增加了附录 VI Expression of Numerals, Mathematic Symbols and Formulas（数字、符号和公式的表达）。第 2 版教材的更新和补充内容全部由何宏完成。

编　者

第1版前言

随着我国信息产业的发展，对外交流日益增多，我国教育部已经要求各高等院校积极推广使用英语等外语进行专业课教学，为培养国际型复合人才奠定基础。而专业英语课作为目前我国大学非英语专业的专业必修课，是培养学生用英语表达科学技术能力的桥梁。

本书是根据普通高等院校本科生专业英语教学大纲的要求，在多年课程教学实践的基础上，针对电气信息类专业编写的专业英语教材。鉴于专业英语课依然是语言类课程，本书为了适应工业发展和社会的需求，着力克服现有专业英语教材阅读文章内含过多的公式、图表，没有汉译英技巧的相关理论和听说内容的缺点。本书完全改变了现有的教材以专业阅读材料为主线的编写模式，采用以语言翻译、写作技巧为主要学习内容，电气信息类的科普文章和专业文献作为辅助内容的方法编写。翻译技巧内容不仅包含英译汉的内容，还包含了汉译英的内容。此外，本书为增强学生专业英语听说的能力，还增加了发音技巧和演讲方法的内容，并配有相应的听说练习。本书共有13个单元介绍英译汉方法，5个单元介绍汉译英方法，2个单元介绍写作技巧，阅读材料涵盖了电子信息工程、通信工程、自动化和电气工程及其自动化四个专业，选材广泛，内容由浅入深。书中包括了英译汉精读和泛读文献各13篇、汉译英练习5个单元，并给出电气信息类各专业领域的常用词汇，教材内容既有针对性，又适用于电气信息类多个专业的专业英语教学，应用范围广，可选择性强，还可供其他科技人员作为英文翻译的参考。

本书由上海师范大学何宏主编，上海师范大学张相芬和华东理工大学黎冰参加编写。全书的英译汉方法（Part I Unit 1 A～Unit 13 A）、发音技巧（Part I Unit 1 D～Unit 6 D）、演讲方法（Part I Unit 7 D～Unit 13 D）和 Part III 汉译英方法由何宏编写。Part I 中前6单元阅读部分（Part I Unit 1 B、C～Unit 6 B、C）和 Part II 写作部分由张相芬编写。Part I 中阅读部分（Part I Unit 7 B、C～Unit 12 B、C）由黎冰编写。Part I Unit 13 B 和 C，以及附录中的自动化、电气工程和计算机专业词汇由何宏编写，附录中的通信工程和电子工程专业词汇由张相芬编写。全书由何宏统稿。此外，华东理工大学的硕士生李程凯和上海师范大学的硕士生刘鑫也参加了本书部分章节的文字输入和校对工作，在此对他们表示诚挚的感谢。

本书有完整的翻译练习答案，并有听说部分的语音材料供大家使用，读者可以登录网站 http：//www.cmpedu.com/main/索取。由于编者水平有限，书中难免有不足之处，欢迎大家批评指正，我们将不胜感激。

编 者

目 录

第 2 版前言
第 1 版前言

PART I　Techniques of EST E-C Translation
（科技英语英译汉翻译技巧）

Unit 1 ··· 1
 A　An Introduction to EST（科技英语基础知识） ·· 1
 B　Circuit Theory（电路理论） ·· 8
 C　Basic Concepts of Artificial Intelligence（人工智能基本概念） ························· 12

Unit 2 ··· 16
 A　Translation of Scientific and Technological Terminology（科技术语的翻译方法） ····· 16
 B　Introduction to Signals and Systems（信号与系统介绍） ···································· 20
 C　On the Nature of Artificial Intelligence（人工智能的特征） ······························ 22

Unit 3 ··· 28
 A　Selecting and Determining the Meaning of a Word（词义的选择和确定） ····· 28
 B　Communication Modeling（通信建模） ·· 32
 C　Artificial Intelligence in Medicine（医疗中的人工智能） ··································· 34

Unit 4 ··· 39
 A　The Extension of the Meaning of a Word or a Phrase（词义的引申） ············· 39
 B　Typical DSP Applications（数字信号处理的典型应用） ···································· 42
 C　Fundamentals of IoT（物联网基础） ··· 45

Unit 5 ··· 50
 A　The Conversion of Parts of Speech（词性的转换） ··· 50
 B　GPS Vehicle Surveillance Equipment Is Here to Help You（全球卫星定位系统汽车监视装置正在帮助你） ····· 53
 C　Adaptive Assistance：Smart Home Nursing（自适应帮助：智能家居护理） ····· 56

Unit 6 ··· 60
 A　Amplification（增词） ·· 60
 B　Introduction to Control Systems（控制系统介绍） ··· 62
 C　IoT Challenges and Future Directions（物联网的挑战与未来发展） ················ 65

Unit 7 ··· 72
 A　Omission（减词） ·· 72
 B　Design of Control Systems（控制系统设计） ··· 75
 C　The Internet of Things in Manufacturing：Benefits, Use Cases and Trends（制造业的

物联网：优点、实例和发展趋势） 77

Unit 8 81
- A The Conversion of the Elements of a Sentence（句子成分的转换） 81
- B Introduction to PID Controllers（比例积分微分控制入门） 83
- C Applications of Social Robots（社交机器人的应用） 86

Unit 9 93
- A Translation of Passive Sentences（被动句的翻译） 93
- B Electric Motors（电动机） 96
- C Data-Mining Concepts（数据挖掘概念） 99

Unit 10 103
- A Translation of Negative Sentences（否定句的翻译） 103
- B Control of Wind Energy Conversion Systems（风能转换系统控制） 107
- C Data-Mining Process（数据挖掘过程） 109

Unit 11 113
- A Translation of Complex Sentences（复杂句的翻译） 113
- B Thermal Power Plant Simulation and Control（热电站仿真与控制） 116
- C Industry 4.0：the Fourth Industrial Revolution（工业 4.0：第四次工业革命） 119

Unit 12 123
- A Translation of Numerals（数的翻译） 123
- B Chasing the Clouds—Distributed Computing and Small Business（追逐云——分布式计算与小型商业） 126
- C Challenges and Benefits of Industry 4.0（工业 4.0 的挑战和优势） 130

PART II Practical Writing Techniques
（实用的写作技巧）

Unit 1 Abstracts（摘要） 138
1. Types of Abstracts（摘要的类型） 138
2. Essential Elements of the Abstract（摘要的基本要素） 139
3. Writing Techniques of the Abstract（摘要撰写技巧） 139
4. Key Words（关键词） 140
5. Sample Abstracts（摘要举例） 141
6. Useful Sentence Structures in English Abstracts（英文摘要常用句式） 142

Unit 2 Resumes（简历） 145
1. Introduction（概述） 145
2. Writing Techniques of the Resume（简历写作技巧） 145
3. Key Points for Resume Writing（简历写作要点） 147
4. Sample Resumes（个人简历例文） 147

PART III Techniques of EST C-E Translation
（科技英语汉译英翻译技巧）

Unit 1　Techniques of Chinese Words Translation I（汉语词汇翻译方法 I）……………… 150

Unit 2　Techniques of Chinese Words Translation II（汉语词汇翻译方法 II）……………… 156

Unit 3　Techniques of Chinese Words Translation III（汉语词汇翻译方法 III）……………… 160

Unit 4　Techniques of Chinese Words Transaltion IV（汉语词汇翻译方法 IV）……………… 162

Unit 5　Translation of Emphatic Sentences（强调句的翻译）………………………………… 166

Unit 6　Translation of Attributive Clauses（定语从句的翻译）……………………………… 168

Unit 7　Translation of Adverbial Clauses（状语从句的翻译）………………………………… 172

Unit 8　Omission in Sentence Translation（句子翻译中的省略）…………………………… 175

附　　录

附录 I　Glossary for Electronic Engineering（电子工程专业技术词汇）…………………… 179

附录 II　Glossary for Communication Engineering（通信工程专业技术词汇）……………… 185

附录 III　Glossary for Automation（自动化专业技术词汇）………………………………… 191

附录 IV　Glossary for Electrical Engineering（电气工程专业技术词汇）…………………… 198

附录 V　Glossary for Computer Science（计算机专业技术词汇）…………………………… 204

附录 VI　Expression of Numerals, Mathematic Symbols and Formulas（数字、符号和公式的表达）……………………………………………………………………… 211

参考文献 ……………………………………………………………………………………… 215

PART I Techniques of EST E-C Translation
（科技英语英译汉翻译技巧）

Unit 1

A An Introduction to EST（科技英语基础知识）

1. The Conception of EST（科技英语的概念）

英语是目前国际上科技领域的主要通用语言。科技英语（English for Science and Technology, EST）主要是科研工作者、技术人员和工程师等交换其专业意见、发明创造、信息数据、实验报告等而使用的语言。为了客观地记录自然现象的发展过程和特点，科技英语在文体上应该精确、简明、严谨，内容上经常包含了数学公式、图表等；措辞上常使用典型的句式及大量的专业或半专业术语，这就使科技英语不同于普通英语（Ordinary English）。

根据内容上所涉及的专业知识的深浅不同，科技英语又可以分成专业科技英语（English for Specialized Science and Technology, ESST）和通俗科技英语（English for Common Science and Technology, ECST）。顾名思义，专业科技英语应用于科技专业领域，使用者一般是各专业领域中的科研技术人员。所以，若没有相关的专业知识，即便母语是英语的读者也不一定完全懂专业科技英语。

例如：Capacitors

Capacitors play a vital role in modern electronics. A capacitor is a device consisting of two conductors separated by vacuum or an insulating material. Capacitors are used in a wide variety of electric circuits and are a vital part of modern electronics. When charges of equal magnitude and opposite sign are placed on the conductors of a capacitor, an electric field is established in the region between them, with a corresponding potential difference between the conductors. The relations among charge, field, and potential can be analyzed by using the results of the two preceding chapters. For a given capacitor, the ratio of charge to potential difference is a constant, called the capacitance. Placing charges on the conductors requires an input of energy; this energy is stored in the capacitor and can be regarded as associated with the electric field in the space between conductors. When this space contains an insulating material (a dielectric) rather than vacuum, the capacitance is increased.

When we speak of a capacitor as having charge Q, we mean that the conductor at higher potential has a charge Q and the conductor at lower potential has a charge $-Q$ (assuming Q is a positive quantity). This interpretation should be kept in mind in the following discussion and examples.

通俗科技英语涉及的是一些普及性的基本科技知识，也就是我们常说的科普文章。事实上，随着科学技术的飞速发展和全球化进程的不断加快，作为科学技术交流媒介的科普著作正起着越来越重要的作用。国际上有大量的期刊和书报刊登通俗的科技类文章。

例如：Computer Engineering

Computer engineering involves the development and application of computer system, which perform tasks, such as mathematical calculations or electronic communication, under the control of a set of instructions called a program. Programs usually reside within the computer and are retrieved and processed by the computer's electronics, and the program results are stored or routed to output devices, such as video display monitors or printers. Computers are used to perform a wide variety of activities with reliability, accuracy, and speed.

2. Basic Features of EST（科技英语的基本特点）

（1）科技英语的文体特点

文体的原始义是指"以文字修饰思想的一种特殊方式"，在古希腊被视为一种语言说服的技巧。而现在文体是指一定的话语秩序所形成的文本体式，它折射出写作者独特的精神结构、体验方式、思维方式和社会历史文化精神。随着科学技术的不断发展，科技英语已形成一种独立的英语文体，它与传统的新闻报刊文体、公文文体、描述文体、叙述文体及应用文体构成了当代六大英语文体。了解科技英语的文体特点有助于提高对科技篇章的理解能力，更快掌握科学技术的发展信息。

科技英语具有普通英语的共性，但也有别于普通英语目的性表述的个性化和特殊性。科技文章中的词、句、章在表述上具有科学性、规范性和简洁性。科学性是指客观而准确地阐述科技问题。规范性是指科技文章的表达方式是程式化的。例如，使用公认的符号和公式，选用不会引起歧义的术语和源于拉丁语和法语的词汇，以及专业科技词汇，从而实现科学的、规范的阐述。简洁性是指使用精练的语言和准确朴实的语言表达。

例如，下面分别用普通英语和科技英语写成的《Natural and Synthetic Rubber》一文，普通英语具有简单、通俗、口语化等特征，相比较而言，科技英语则更加严肃客观，而且专业性强。

Natural and Synthetic Rubber	
普通英语	科技英语
People **get** natural rubber from rubber trees as a white, milky liquid, **which is called** latex, they **mix it with** acid, and dry it, **and then they send it** to countries all over the world. As the rubber industry **grew**, people **needed** more and more rubber. They **started** rubber plantations in countries with hot, **wet weather conditions**, but these still could not **give enough** raw rubber **to meet the needs** of growing industry. It was **not satisfactory** for industry to depend on supplies, **which come** from so far away from the industrial areas of Europe. It was always possible that wars or shipping trouble could stop supplies. For many years people **tried to make something to take its place**, but they **could not do it. In the end**, they **found a way** of making **artificial, man-made** rubber which is in many ways **better than** and in some ways **not as good as** natural rubber. They make artificial, man-made rubber in factories by a complicated chemical process. It is usually cheaper than natural rubber. **Today, the world needs so much rubber** that we use both natural and artificial rubber in large **amounts**.	Natural rubber **is obtained** from rubber trees as a white, milky liquid **known as** latex. This is **treated with** acid and dried **before being dispatched** to countries all over the world. As the rubber industry **developed**, more and more rubber was **required**. Rubber plantations were **established** in countries with a hot, **humid climate**, but these still could not **supply sufficient** raw rubber to **satisfy the requirements** of developing industry. It was **unsatisfactory** for industry to depend on supplies **coming** from so far away from the industrial areas of Europe. It was always possible that supplies could be stopped by wars or shipping trouble. For many years, **attempts were made to produce a substitute**, but they **were unsuccessful. Finally, a method was discovered** of producing **synthetic** rubber which is in many ways **superior** and in some ways **inferior to** natural rubber. Synthetic rubber is produced in factories by complicated chemical process. It is usually cheaper than natural rubber. **At present, the world requirements for rubber are so great** that both natural and synthetic rubber are used in **quantities**.

PART I　Techniques of EST E-C Translation（科技英语英译汉翻译技巧）

(2) 科技英语的句法和修辞的特点

目前，科技英语已经发展成为一种重要的英语文体。科技文体崇尚严谨周密，概念准确，逻辑性强，行文简练，重点突出，句式严整，少有变化，常用前置性陈述，即在句中将主要信息尽量前置，通过主语传递主要信息。与普通英语相比，科技英语具有自己突出的特点。

1) 广泛使用被动语句。科技英语叙述的对象往往是事物、现象或过程，其注重的是其叙述的客观事实，强调的是所叙述的事物本身，而并不需要过多注意它的行为主体是什么。英语中的被动语态不仅比较客观，而且可使读者的注意力集中在所叙述的客体上。根据英国利兹大学约翰·斯韦尔斯（John Swales）的统计，科技英语中的谓语至少三分之一是被动态。这是因为第一、二人称使用过多，会造成主观臆断的印象。因此尽量使用第三人称叙述，采用被动语态。例如：

Example 1：Attention must be paid to the working temperature of the machine.

译文：应当注意机器的工作温度。

而很少说：You must pay attention to the working temperature of the machine.

译文：你们必须注意机器的工作温度。

此外，如前所述，科技文章将主要信息前置，放在主语部分。这也是广泛使用被动态的主要原因。又如：

Example 2：This steel alloy is believed to be the best available here.

译文：人们认为这种合金钢是这里能提供的最好的合金钢。

Example 3：Computers may be classified as analog and digital.

译文：计算机可分为模拟计算机和数字计算机两种。

Example 4：The switching time of the new-type transistor is shortened three times.

译文：新型晶体管的开关时间缩短了三分之二。（或"缩短为三分之一"）

Example 5：The temperature of the liquid is raised by the application of heat.

译文：加热可以提高液体温度。

Example 6：Useful facts may be collected either by making careful observation or by setting up experiment.

译文：通过仔细地观察或做实验可以收集到有用的数据。

例 5 和例 6 中的 by 短语表达的不是行为的发出者，而是方式、方法或手段。而在一般英语中，by 短语大多数表达的是行为的发出者。

Example 7：It seems that these two branches of science are mutually dependent and interacting.

译文：看来这两个科学分支是相互依存、相互作用的。

Example 8：It has been proved that induced voltage causes a current to flow in opposition to the force producing it.

译文：已经证明，感应电压使电流的方向与产生电流的磁场力方向相反。

2) 长句多，且句式变化少。在科技英语中表示某些复杂概念时用的长复合句大大多于一般英语。长句的特点是从句和短语多，同时兼有并列结构或省略、倒装语序，结构显得复杂。但长句所表达的科技内容严密性、准确性和逻辑性较强，这也是长句在科技英语中常见的主要原因。从下面两个例句可以看出科技英语长句的特点：

Example 1：With the advent of the space shuttle, it will be possible to put an orbiting solar power plant in stationary orbit 24,000 miles from the earth that would collect solar energy almost continuously and convert this energy either directly to electricity via photovoltaic cells or indirectly with flat-plate or focused collectors that would boil a carrying medium to produce steam that would drive a turbine that then in turn would generate electricity.

译文：随着航天飞机的出现，有可能把一个沿轨道运行的太阳能发电站送到离地球24 000mile（1mile≈1.609km）的一条定常轨道上去。这个太阳能发电站几乎不间断地取用太阳能。它还能够用光电池将太阳能直接转换成电能，或者用平板集热器或聚焦集热器将太阳能间接转换成电能，即集热器使热传导体汽化，驱动涡轮机发电。

上例中 solar power plant 带有一个距离较远的定语从句，该从句中又含有另外三个定语从句。这四个定语从句均由 that 引出，环环相套，层见叠出。尽管例句句子结构复杂，但关系清楚，逻辑性强。

Example 2：Only by studying such cases of human intelligence with all the details and by comparing the results of exact investigation with the solutions of AI (artificial intelligence) usually given in the elementary books on computer science can a computer engineer acquire a thorough understanding of theory and method in AI, develop intelligent computer programs that work in a human-like way, and apply them to solving more complex and difficult problems that present computer can't.

译文：只有很详细地研究这些人类智能情况，并把实际研究得出的结果与基础计算机科学书上给出的人工智能结论相比较，计算机工程师才能彻底地了解人工智能的理论和方法，开发出具有人类智能的计算机程序，并将其用于解决目前计算机不能解决的更复杂和更难的问题。

本句为复合句，一主一从。主句有一个主语，三个并列谓语。句子以"only + 状语"开头，主句主、谓语部分倒装，主语 a computer engineer 处于谓语之间，形成 can a computer engineer acquire, develop and apply 这样一种语序。过去分词 given 引导的短语做后置定语，修饰前面的 solutions of AI。另外，长状语 only... computer science 修饰主句谓语 can acquire, develop and apply。that 引导了一个后置定语从句，修饰 more complex and difficult problems。

3）非谓语动词多。如前所述，科技文章要求行文简练，结构紧凑，为此，往往使用分词短语（participle phrase）代替定语从句（attributive subordinate clause）或状语从句（adverbial subordinate clause）；使用分词独立结构代替状语从句或并列分句；使用不定式短语（infinitive phrase）代替各种从句；介词和动名词（gerund）短语代替定语从句或状语从句。这样可缩短句子，又比较醒目。试仔细阅读下列各例句：

Example 1：A direct current is a current *flowing always* in the same direction.

译文：直流电是一种总是沿同一方向流动的电流。

Example 2：*Radiating from the earth*, heat causes air currents to rise.

译文：热量由地球辐射出来时，使得气流上升。

Example 3：A body can move uniformly and in a straight line, *there being* no cause to change that motion.

译文：如果没有改变物体运动的原因，那么物体将做匀速直线运动。

Example 4：Vibrating objects produce sound waves, *each vibration producing one sound wave.*

译文：振动着的物体产生声波，每一次振动产生一个声波。

Example 5：Materials *to be used* for structural purposes are chosen so as to behave elastically in the environmental conditions.

译文：结构材料的选择应使其在外界条件中保持其弹性。

Example 6：There are different ways *of changing energy* from one form into another.

译文：将能量从一种形式转变成另一种形式有各种不同的方法。

Example 7：*In making the radio waves* correspond to each sound in turn, messages are carried from a broadcasting station to a receiving set.

译文：使无线电波依次对每一个声音做出相应变化时，信息就由广播电台传递到接收机。

4）时态运用有限。科技英语所运用的时态大都限于一般现在时、一般过去时、现在完成时、过去完成时和一般将来时这几种，其他时态运用很少。其中一般现在时给人以"无时间性"的概念，以排除任何与时间关联的误解，主要应用于对定义、定理、公式或图表进行科学解说，或者用于表述一些通常发生或并非时限的自然现象、过程和规则等。

Example 1：Common salt dissolves in water.

译文：食盐溶于水。

Example 2：Figure 2 shows the principal layout of an oil refinery.

译文：图2显示了炼油厂的总布置图。

一般过去时在科技英语中常用于叙述过去进行的研究情况。例如，一个描述过去试验情况的句子：

Example 3：Rice grew better, under the other conditions of these tests, when ammonium sulphate was added to the soil.

译文：在这些试验的其他条件不变的情况下，当土壤里添加了硫酸铵时，稻子生长得较好。

若所描述事物与现在相关且影响较大，则用现在完成时。

Example 4：The reaction has already come to the end.

译文：反应已经终止。

Example 5：One of the most striking characteristics of modern science has been the increasing trend towards closer cooperation between scientists and scientific institutions all over the world.

译文：现代科学的最显著特点之一，就是全世界科学家及科学机构之间不断发展更为密切的合作趋势。

而在介绍过去曾经做过的工作时，用过去完成时。

Example 6：The data had no sooner been charted than analysis was started.

译文：资料刚在图上填完，分析就开始了。

当讨论进行中的项目研究时，科技英语常采用一般将来时，说明未来拟定的活动。

Example 7：The scientists and technicians will carry out a very important test next month.

译文：科学家和技术人员下个月将进行一项非常重要的测试。

5）修辞手法单调。英语中通常有夸张（hyperbole）、明喻（metaphor）、借喻（metonymy）、拟人（personification）和对照（contrast）等修辞手法。这些手法在英语的文学文体中是常见的，但在科技英语中却很少见。这是因为科技英语注重叙述事实和逻辑推理，若采用文学上的修辞法，会破坏科学的严肃性，反而弄巧成拙。

3. Principles of EST Translation（翻译的标准）

翻译标准是指导翻译的准绳，也是衡量译文质量的尺度。我国清末翻译家严复第一个提出了"信、达、雅"（fidelity, coherence, elegance）的翻译标准。即"信"是指"忠实"，"达"是指"流畅"，"雅"是指"尔雅"。随着科学技术的不断进步和翻译工作的不断发展，翻译标准也发生了一些新的变化，人们对不同类型的文章和作品提出了不同的要求和标准。但是"信、达、雅"的提法严谨准确，历来为后人推崇，至今仍有很大影响，因此科技英语的翻译依然要求符合这三个基本准则。

下面就根据这三个标准，结合电信类科技英语的特点，谈谈如何做到"信、达、雅"。

（1）如何做到"信"

"信"是翻译标准中的第一个标准，也是最重要、最基本的标准。无论是文学翻译还是科技翻译，译文要忠于原文已成为翻译界的共识。科技文体的翻译，译文要做到"信"，译者必须具备以下几点要求：

1）要有丰富的专业知识。翻译国外科学教材和文献自然要有丰富的专业科技知识，这是对译者的基本要求之一。也就是说，译者在这一领域必须是专业人士，因为没有专业知识就很难做到正确理解原文内容。此外，电信类学科包含了电子、通信、自动化、电气专业，所涉及的领域十分广泛。此外，译者还需要具备扎实的数学、物理学和化学等学科的知识，否则翻译的专业术语名称会不恰当，甚至造成误译。

Example: This apparently simple function of the transformer makes it as vital to modern industry as the gear train which, as a "transformer" of speed and torque represents an interesting analogy.

译文：变压器的这种显而易见的基本功能，对现代工业起着重要作用。这种作用可以与齿轮对近代工业的巨大作用做一有趣的比较，齿轮是机械转速与力矩的"转换器"。

transformer 原义是指"使变化的人或事物"，专业上有"变压器，变换器，互感器"的含义，还有电影里出现的"变形金刚"也是该词。但是在这句话里，第一个 transformer 是指电气学科里的"变压器"，第二个 transformer 则是对变压器与齿轮进行功能比较，指两者都具有转换功能。若在翻译时没有一定的电学知识和相关的机械领域知识，这句话很容易译错。

2）要遵守约定俗成的原则。目前，电信类学科中的许多专业术语早已被译成汉语，这些汉译名称要么为学术界公认，要么为广大读者所认可，即已经约定俗成。译者必须熟悉这些专业术语的汉译名称，在翻译中不可自己创造，以免读者误解。

3）要避免误译、胡译、漏译。与其他文体的翻译一样，译者翻译国外科技文献时也要有认真负责的态度，不能望文生义，或者由于原文结构复杂、句子较长而造成误译、胡译甚至漏译。

（2）如何做到"达"

"达"的目的就是使译文通顺流畅，符合汉语的表达方式。译者在翻译国外科技文献时

PART I Techniques of EST E-C Translation（科技英语英译汉翻译技巧）

要做到"达"必须符合以下几点要求：

1）尽量使用直译。所谓直译就是既忠实于原文的内容，又忠实于原文形式的翻译方式。从语言的语法结构来看，世界上的各种语言大致可分为综合性语言和分析性语言两大类。英语的词形变化很少，主要依靠词序，或者通过虚词和某些语言习惯来表示各语法成分之间的关系，基本上属于分析性语言。而汉语根本就没有词形的变化，可以说是纯粹的分析性语言。正是英汉两种语言之间的这个共性，为英汉之间的直译提供了很多便利。所以，科技文体翻译中尽量采用直译，可以体现原文的科技内容。

Example：In some automated plants electronic computers control the entire production line.

译文：在某些自动化工厂，电子计算机控制着整个生产线。

2）不能直译的才采用意译。所谓意译，就是只忠实于原文的内容，不拘泥于原文的形式的一种翻译方法。由于英语基本上属于分析性语言，所以句子结构不够严谨，尤其是在一些复杂的长句中，句子各成分之间以及各分句之间的相互关系不够明确，有时不太容易从语法结构上加以辨别，往往需要借助句子的具体含义来进行分析，这使我们在翻译过程中遇到许多困难。我们可考虑意译的方法加以解决。在科技英语翻译中直译为主，意译为辅助翻译方法。

Example 1：Computer viruses are programs designed to replicate and spread, sometimes *without indicating that they exist*.

译文：计算机病毒是可以复制和传播的程序，有时它们的存在并<u>不为人所知</u>。

Example 2：The year, 1998, *saw* great development in the field of computer.

译文：1998年计算机领域<u>有了</u>很大的发展。

Example 3：Control systems frequently *employ* components of many types.

译文：控制系统频繁<u>使用</u>各种类型的元件。

Example 4：Automatic control systems are physical systems which have *dynamic behavior*.

译文：自动控制系统是具有<u>动态特性</u>的物理系统。

Example 5：The differences between single-board microcomputer and single chip microcomputer *do not stop here*.

译文：单板机和单片机的差别<u>不仅如此</u>。

Example 6：Amplification means the transformation of little currents into big ones, without *distortion* of the shape of current fluctuation.

译文：所谓放大就是把小电流变为大电流，而又不使电流波形<u>失真</u>。

（3）如何做到"雅"

译文要达到"雅"这个标准，译者应该考虑到英语与汉语存在着巨大的文化背景差异，使它们的表达方式截然不同，这时我们往往要改换表达的形式进行翻译。并且翻译时要避免形式主义，切忌逐字逐句对等翻译，生搬硬套地硬译只会出错，甚至在闹笑话。

请翻译以下例句，并对照参考译文体会"雅"的含义：

Example 1：Control Center *Smoking Free*.

译文：控制中心，<u>严禁吸烟</u>。

Example 2：The importance of computer in the use of automatic control *cannot be overestimated*.

译文：计算机在自动控制应用上的重要性<u>怎么估计也不会高</u>。

Example 3: Continuous control *readily leads itself to understanding of* feedback control theory using relatively uncomplicated mathematics.

译文：由于使用的数学相对简单，故连续控制<u>有利于反馈控制理论的理解</u>。

Example 4: This machine is *the last word in technical skill.*

译文：这台机器<u>在技术上达到了领先水平</u>。

4. Process of EST Translation（科技英语翻译过程）

科技英语的翻译过程是正确理解原文和创造性地用汉语再现原文的过程，大体上可以分为理解、表达和校核三个阶段：

1）理解过程。对原文做透彻的理解是准确翻译的基础和关键，这不仅包括理解原文的专业术语、句法结构、逻辑关系，还包括理解原文的专业表达方式。理解过程主要通过专业领域知识和原文的上下文来探求正确的译法。有时句子的含义不能够看词语的表面意思，而应当仔细推敲，理解写作者所要表达的真正含义。

Example: Scientific discoveries and inventions do not always influence the language in proportion to their importance.

初译：科学发现和发明就其重要性的比例而言，并不一定对语言有什么影响。

校核后译文：科学发现和发明对语言的影响，并不取决于其重要性。

2）表达过程。表达过程就是要把译者从原文理解的内容用汉语重新表达出来。表达的好坏主要取决于对原文理解的深度以及汉语语言的修养程度。表达是理解的结果，但是理解正确并不意味着必然能正确表达。比如下面两句就是能够看懂却不好表达的句子：

Example 1: A translator has to know everything of something and something of everything.

初译：翻译人员对一些事情要什么都懂，对所有事情又要多少懂一些。

应译为：翻译人员既要学问精深，又要知识广博。

Example 2: One may as well not know a thing at all as know it but imperfectly.

译文：与其一知半解，不如全然不知。

3）校核过程。校核是理解与表达的进一步深化，是对原文内容进一步核实以及对译文语言进一步推敲的过程。翻译时尽管十分细心，但译文中仍然会有错、漏、别字或字句欠妥的地方。因此，译文的校核过程是在翻译中必不可少的环节，通过校核能够使译文更加忠实于原文，语句更加通顺。比如下面的两个短句：

Example 1: This possibility was supported to a limited extend in the tests.

初译：在实验中，这一可能性在一定程度上被支持了。

校核后译文：这种可能性在实验中在一定程度上得到了证实。

Example 2: There are crossed transverse steady and longitudinal alternating fields.

初译：有交叉横向稳定和纵向交变磁场。

校核后译文：在纵向交变磁场上叠加了一个横向稳定磁场。

B Circuit Theory（电路理论）

1. The Electrical Circuit

An electrical circuit or electrical network is an array of interconnected elements wired so as to be capable of conducting current.[1] The fundamental two-terminal elements of an electrical circuit are the resistor, the capacitor, the inductor, the voltage source, and the current source. The value of a resistor is known as its resistance R, and its units are ohms (Ω). For the capacitor, the capacitance C, has units of farads (F), the value of an inductor is its inductance L, the units of which are henries (H). In the case of the voltage sources, a constant, time invariant source of voltage, or battery, is distinguished from a voltage source that varies with time. The latter type of voltage source is often referred to as a time varying signal or simply, a signal. In either case, the value of the battery voltage E, and the time varying signal $v(t)$, is in units of volts (V). Finally, the current source has a value I, in units of amperes (A), which is typically abbreviated as amps.

Elements having three, four, or more than four terminals can also appear in practical electrical networks. The discrete component bipolar junction transistor (BJT), is an example of a three-terminal element, in which the three terminals are the collector, the base, and the emitter. On the other hand, the monolithic metal-oxide-semiconductor field-effect transistor (MOSFET) has four terminals: the drain, the gate, the source, and the bulk substrate.

Multi-terminal elements appearing in circuits identified for systematic mathematical analyses are routinely represented, or modeled, by equivalent sub-circuits formed of only interconnected two-terminal elements.[2]

2. Current, Voltage, Power

The current flow through an element that is capable of current conduction is the time rate of change of the transferred charge. The unit of charge is the coulomb, time t is measured in seconds, and the resultant current is measured in units of amperes.

The terminal voltage, $v(t)$, corresponding to the energy, $w(t)$, required to transfer an amount of charge, $q(t)$, across an arbitrary cross-section of the element is: $v(t) = \dfrac{dw(t)}{dq(t)}$. Here, $v(t)$ is in units of volts when $q(t)$ is expressed in coulombs, and $w(t)$ is specified in joules.

The time rate of change of the applied energy is the power, which is in units of watts. In electrical circuits, the power delivered to an element is simply the product of the voltage applied across the terminals of the element and the resultant current conducted by that element.

3. Circuit Classifications

Electrical elements and circuits in which they are embedded are generally codified as linear or nonlinear, active or passive, time varying or time invariant, and lumped or distributed.

1) Linear or nonlinear circuits. A linear two-terminal circuit element is one for which the voltage developed across, and the current flowing through, are related to one another by a linear algebraic or a linear integro-differential equation. If the relationship between terminal voltage and corresponding current is nonlinear, the element is said to be nonlinear. A linear circuit contains only linear circuit elements, while a circuit is said to be nonlinear if at least one of its embedded electrical elements is nonlinear.[3]

2) Active or passive circuits. An electrical element or network is said to be passive if the power delivered to it, is positive. In contrast, an element or network to which the delivered power is

negative is said to be active; that is, an active element or network generates power instead of dissipating it. Conventional two-terminal resistors, capacitors, and inductors are passive elements. It follows that networks formed of interconnected two-terminal resistors, capacitors, and inductors are passive networks. Two-terminal voltage and current sources generally behave as active elements. Multi-terminal configurations, whose models exploit dependent sources, can behave as either passive or active networks.

3) Time varying or time invariant circuits. The elements of a circuit are defined electrically by an identifying parameter, such as resistance, capacitance, inductance, and the gain factors associated with dependent voltage or current sources. An element whose identifying parameter changes as a function of time is said to be a time varying element. If said parameter is a constant over time, the element in question is time invariant. A network containing at least one time varying electrical element it is said to be a time varying network. Otherwise, the network is time invariant. Excluded from the list of elements whose electrical character establishes the time variance or time invariance of a considered network are externally applied voltage and current sources. Thus, for example, a network with internal elements that are exclusively time-invariant resistors, capacitors, inductors, and dependent sources, but which is excited by a sinusoidal signal source, is nonetheless a time-invariant network.

4) Lumped or distributed circuits. Electrons in conventional conductive elements are not transported instantaneously across elemental cross sections, but their transport velocities are very high. In fact, these velocities approach the speed of light. Electrons and holes in semiconductors are transported at somewhat slower speeds, but generally no less than an order of magnitude or so smaller than the speed of light. The time required to transport charge from one terminal of a two-terminal electrical element to its other terminal, compared with the time required to propagate energy uniformly through the element, determines whether an element is lumped or distributed.[4] In particular, if the time required to transport charge through an element is significantly smaller than the time required to propagate the energy through the element that is required to incur such charge transport, the element in question is said to be lumped. On the other hand, if the charge transport time is comparable to the energy propagation time, the element is said to be distributed.

4. Kirchhoff's Circuit Laws

Kirchhoff's Current Law (KCL) is one of two fundamental laws in electrical engineering, the other being Kirchhoff's Current Law (KCL). It states that the algebraic sum of currents in a network of conductors meeting at a point is identically zero at all instants of time.

Kirchhoff's Voltage Law (or Kirchhoff's Loop Rule, KVL) is a result of the electrostatic field being conservative. It states that the total voltage around a closed loop is identically zero at all instants of time.

5. Network Theorems

1) Superposition theorem. The superposition theorem for electrical circuits states that the response (Voltage or Current) in any branch of a bilateral linear circuit having more than one independent source equals the algebraic sum of the responses caused by each independent source

PART I   Techniques of EST E-C Translation（科技英语英译汉翻译技巧）   11

acting alone, while all other independent sources are replaced by their internal impedances.

2) Thevenin's theorem. It states that any combination of voltage sources, current sources, and resistors with two terminals is electrically equivalent to a single voltage source V and a single series resistor R.

3) Norton's theorem. Norton's theorem is an extension of Thévenin's theorem and was introduced in 1926 separately by two people: Hans Ferdinand Mayer and Edward Lawry Norton. It states that any collection of voltage sources, current sources, and resistors with two terminals is electrically equivalent to an ideal current source I, in parallel with a single resistor R.

New Words and Expressions

two-terminal element   二端组件
resistor   n.   电阻，电阻器
capacitor   n.   电容，电容器
inductor   n.   电感，电感器
voltage source   电压源
resistance   n.   阻力，电阻
capacitance   n.   电容，电流容量
inductance   n.   电感，感应系数，自感应
Ohm   n.   欧姆（电阻单位）
Farad   n.   法拉（电容单位）
Henry   n.   亨利（电感单位）
Volt   n.   伏特，电压
Ampere   n.   安培（计算电流的标准单位）
bipolar junction transistor   双极面结型晶体管
collector   n.   集电极

base   n.   基极
emitter   n.   发射极
monolithic   n.   单集成电路，单片电路
active (passive) circuit   有（无）源电路
(non) linear circuit   （非）线性电路
time invariant circuit   时不变电路
lumped (distributed) circuit   集总（分散）参数电路
superposition theorem   叠加定理
Thevenin's theorem   戴维南定理
Norton's theorem   诺顿定理
metal-oxide-semiconductor field-effect transistor   金属氧化物半导体场效应晶体管
Kirchhoff's current (voltage) law   基尔霍夫电流（压）定律

Notes

[1] An electrical circuit or electrical network is an array of interconnected elements wired so as to be capable of conducting current. 电路或者电网络由一系列电子元件组成，这些元件通过导线连接起来以传导电流。

[2] Multi-terminal elements appearing in circuits identified for systematic mathematical analyses are routinely represented, or modeled, by equivalent sub-circuits formed of only interconnected two-terminal elements. 为进行系统数学分析，电路中的多端元件一般用二端元件组成的子电路等效表示。

[3] A linear circuit contains only linear circuit elements, while a circuit is said to be nonlinear if at least one of its embedded electrical elements is nonlinear. 线性电路仅仅包含线性电路元件，而至少包含一个非线性电子元件的电路才称为非线性电路。

[4] The time required to transport charge from one terminal of a two-terminal electrical element to

its other terminal, compared with the time required to propagate energy uniformly through the element, determines whether an element is lumped or distributed. 二端元件中电荷从元件一端转移到另一端所需时间和相应能量传输时间的相对关系决定了该元件是集总参数元件还是分散参数元件。

C Basic Concepts of Artificial Intelligence（人工智能基本概念）

Since the invention of computers or machines, their capability to perform various tasks went on growing exponentially. Humans have developed the power of computer systems in terms of their diverse working domains, their increasing speed, and reducing size with respect to time. While exploiting the power of the computer systems, the curiosity of human, lead him to wonder, "Can a machine think and behave like humans do?" Thus, the development of Artificial Intelligence (AI) started with the intention of creating similar intelligence in machines that we find and regard high in humans.

1. What Is Artificial Intelligence?

According to the father of AI, John McCarthy, it is "The science and engineering of making intelligent machines, especially intelligent computer programs". AI is a way of making a computer, a computer-controlled robot, or a software think intelligently, in the similar manner the intelligent humans think. AI is accomplished by studying how human brain thinks, and how humans learn, decide, and work while trying to solve a problem, and then using the outcomes of this study as a basis of developing intelligent software and systems. It is the science and engineering of making intelligent machines, especially intelligent computer programs. It is related to the similar task of using computers to understand human intelligence, but AI does not have to confine itself to methods that are biologically observable. Incomputer science, the field of AI research defines itself as the study of "intelligent agents": any device that perceives its environment and takes actions that maximize its chance of success at some goal. [1] Colloquially, the term "artificial intelligence" is applied when a machine mimics "cognitive" functions that humans associate with other human minds, such as "learning" and "problem solving". The overall research goal of AI is to create technology that allows computers and machines to function in an intelligent manner. It may create expert systems which exhibit intelligent behavior, learn, demonstrate, explain, and advice its users, or creating systems that understand, think, learn, and behave like humans.

（1）Intelligence and Its Types

Intelligence is the ability of a system to calculate, reason, perceive relationships and analogies, learn from experience, store and retrieve information from memory, solve problems, comprehend complex ideas, use natural language fluently, classify, generalize, and adapt new situations. Varying kinds and degrees of intelligence occur in people, many animals and some machines. As described by Howard Gardner, an American developmental psychologist, the intelligence comes in multifold. You can say a machine or a system is artificially intelligent when it is equipped with at least one and at most all intelligences in the following:

1) Linguistic intelligence. The ability to speak, recognize, and use mechanisms of phonology (speech sounds), syntax (grammar), and semantics (meaning).

2) Musical intelligence. The ability to create, communicate with, and understand meanings made of sound, understanding of pitch, rhythm.

3) Logical-mathematical intelligence. The ability of use and understand relationships in the absence of action or objects. Understanding complex and abstract ideas.

4) Spatial intelligence. The ability to perceive visual or spatial information, change it, and recreate visual images without reference to the objects, construct 3D images, and to move and rotate them. [2]

5) Bodily-Kinesthetic intelligence. The ability to use complete or part of the body to solve problems or fashion products, control over fine and coarse motor skills, and manipulate the objects.

6) Intra-personal intelligence. The ability to distinguish among one's own feelings, intentions, and motivations.

7) Interpersonal intelligence. The ability to recognize and make distinctions among other people's feelings, beliefs, and intentions.

Generally, intelligence is composed of reasoning, learning, problem solving, perception and linguistic intelligence. Reasoning is the set of processes that enables us to provide basis for judgement, making decisions, and prediction. There are broadly two types: inductive reasoning and deductive reasoning. Inductive reasoning conducts specific observations to makes broad general statements, while deductive reasoning starts with a general statement and examines the possibilities to reach a specific, logical conclusion. [3] It also means that if something is true of a class of things in general, it is also true for all members of that class. However, even if all of the premises are true in a statement, inductive reasoning allows for the conclusion to be false.

Learning is the activity of gaining knowledge or skill by studying, practising, being taught, or experiencing something. Learning enhances the awareness of the subjects of the study. The ability of learning is possessed by humans, some animals, and AI-enabled systems. Learning is categorized as follows:

1) Auditory learning. It is learning by listening and hearing. For example, students listening to recorded audio lectures.

2) Episodic learning. To learn by remembering sequences of events that one has witnessed or experienced. This is linear and orderly.

3) Motor learning. It is learning by precise movement of muscles. For example, picking objects, writing, etc.

4) Observational learning. To learn by watching and imitating others. For example, child tries to learn by mimicking her parents.

5) Perceptual learning. It is learning to recognize stimuli that one has seen before. For example, identifying and classifying objects and situations.

6) Relational learning. It involves learning to differentiate among various stimuli on the basis of relational properties, rather than absolute properties. For Example, Adding "little less" salt at the

time of cooking potatoes that came up salty last time, when cooked with adding a tablespoon of salt.

7) Spatial learning. It is learning through visual stimuli such as images, colors, maps, etc. For example, a person can create roadmap in mind before actually following the road.

8) Stimulus-response learning. It is learning to perform a particular behavior when a certain stimulus is present. For example, a dog raises its ear on hearing doorbell.

Problem Solving is the process in which one perceives and tries to arrive at a desired solution from a present situation by taking some path, which is blocked by known or unknown hurdles.[4] Problem solving also includes decision making, which is the process of selecting the best suitable alternative out of multiple alternatives to reach the desired goal are available.

Perception is the process of acquiring, interpreting, selecting, and organizing sensory information. Perception presumes sensing. In humans, perception is aided by sensory organs. In the domain of AI, perception mechanism puts the data acquired by the sensors together in a meaningful manner.

Linguistic intelligence is one's ability to use, comprehend, speak, and write the verbal and written language. It is important in interpersonal communication.

(2) Difference between Human and Machine Intelligence

Human intelligence is something natural, no artificiality is involved in it. In all fields, intelligence is something differently perceived and differently acquired. More specifically, human intelligence is something related to the adaption of various other cognitive process in order to have specific environment. In human intelligence, the word "intelligence" plays a vital role because intelligence is with them all it's need to cogitate and make a step by step plan for performing certain task.

However, AI is designed to add human like qualities in robotic machines. Its major function is to make the robots a good mimic of human beings. In short we can say that it's basically working to make robots a good of copier of humans. Researchers are all time busy in making up a mind that can behave like a human mind, they are putting efforts in doing this task nowadays. Weak AI is the thinking focused towards the development of technology gifted of carrying out pre-planned moves based on some rules and applying these to achieve a certain goal.[5] Strong AI is emerging technology that can think and function same as like humans, not just imitating human behavior in a certain area.

By combining both definitions from the tunnel of technology we can say that human intelligence works naturally and make up a certain thought by adding different cognitive processes. On the other hand, AI is on the way to make up a model that can behave like humans which seems impossible because nothing can replace a natural thing into an artificial thing. Although AI pursues creating the computers or machines as intelligent as human beings, there are still some major differences surely between them:

- Human intelligence is analogue as work in the form of signals and AI is digital, they majorly works in the form of numbers.
- Humans perceive by patterns whereas the machines perceive by set of rules and data.
- Humans store and recall information by patterns, machines do it by searching algorithms. For example, the number 40404040 is easy to remember, store, and recall as its pattern

- is simple.
- Humans can figure out the complete object even if some part of it is missing or distorted; whereas the machines cannot do it correctly.

New Words and Expressions

AI (artificial intelligence)	人工智能		pitch *v.*	投掷
exponentially *adv.*	指数增长地		coarse *adj.*	粗糙的
confine *v.*	强迫		property *n.*	特性，性能
colloquially *adv.*	通俗地说		hurdle *n.*	障碍
multifold *n.*	多种模式		sensory *adj.*	感觉的
mechanism *n.*	机械装置		tunnel *n.*	隧道
syntax *n.*	语法		distorted *adj.*	扭曲的，歪曲的

Notes

［1］In computer science, the field of AI research defines itself as the study of "intelligent agents": any device that perceives its environment and takes actions that maximize its chance of success at some goal. 在计算机科学中，人工智能研究领域定义其为"智能代理"的研究：能感知环境并采取行动使其在某个目标上最大可能地获取成功的任何机制。

［2］Spatial intelligence. The ability to perceive visual or spatial information, change it, and re-create visual images without reference to the objects, construct 3D images, and to move and rotate them. 空间智能是指感知视觉和空间信息，不用参考对象就可改变并重新创造视觉图像，构建三维图像，以及移动和旋转图像的能力。

［3］Inductive reasoning conducts specific observations to makes broad general statements, while deductive reasoning starts with a general statement and examines the possibilities to reach a specific, logical conclusion. 归纳推理是通过具体的观察，做出广泛的一般性结论，而演绎推理则是从一个一般性结论开始，检查可能性是否属于一个具体的、合乎逻辑的结论。

［4］Problem solving is the process in which one perceives and tries to arrive at a desired solution from a present situation by taking some path, which is blocked by known or unknown hurdles. 解决问题的过程是一个人感知环境，并通过存在已知的或未知障碍的路径，试图从当前的情况中得到理想解决方案的过程。

［5］Weak AI is the thinking focused towards the development of technology gifted of carrying out pre-planned moves based on some rules and applying these to achieve a certain goal. 弱人工智能是一种思维方式，这种方式主要研究如何实现基于规则的预定动作的技术，并应用这些技术实现一定的目标。

Unit 2

A Translation of Scientific and Technological Terminology（科技术语的翻译方法）

科技英语中含有大量的专业技术词汇，而不同的学科也有自己的专业词库，这些专业词汇增加了科技文献阅读和翻译的难度。为了准确地翻译专业技术词汇，首先要对科技英语的词汇特点有所了解，并掌握基本的科技术语的翻译方法。

1. Features of Scientific and Technological Terminology（科技术语的特点）

科技文献中除了普通词汇（common word）以外，还有半技术词汇（semi-technical word）和专业术语（terminology）。其中半技术词汇除了具有普通英语中的非技术含义以外，在不同的专业领域其含义也不尽相同。例如：

power 一词在数学中意为"幂，乘方"；在物理中意为"功率"；在光学中意为"放大率，焦强"；在机械学中意为"（杠杆等）机械工具"；在统计学中意为"功效"；在电气工程中意为"电力，动力"。transmission 一词在电气工程中意为"输送，传送"；在信息工程中意为"发射，传播"；在机械学中意为"传动，变速"；在物理中意为"透射"；在医学中意为"遗传"。

相比之下，专业术语的语义精确，所指范围小。例如：bandwidth（带宽），tweeter（高音用扩音器），maglev（磁悬浮列车），diode（二极管）等。这些词汇有的比较难记，但是由于专业术语一般含义稳定，其含义不会因语境的变化而改变。所以，掌握专业术语构词特点，一旦记住其含义，就很难忘却和混淆。

通常掌握了专业术语的构词方法将有助于单词的记忆，科技英语中的专业术语常采用以下的构词方法：

（1）复合词

复合词 Compounding 是用合成法把两个或两个以上的词按照一定的程序排列构成的新词。在整个英语发展过程中，用合成法构成的复合词起着积极的作用，为科技英语提供了大量新词汇。复合词的构词成分可由各种词类组成，且其组合不受英语句法在次序排列上的限制，比较灵活、机动，复合词的构词多半由基本词汇提供。大量使用复合词与缩略词是科技文章的特点之一，复合词从过去的双词组合发展到多词组合，有的词有连字符为分写式，有的无连字符为合写式。例如：

复合名词：barcode（条形码），toothpick（牙签），acidcloud（酸云），airtunnel（空中隧道），pocketoffice（手提式办公设备），solarpond（太阳能池）。

复合形容词：full-enclosed（全封闭的），nuclear-weapon-free（无核武器的），on-and-off-the-road（路面越野两用的）。

复合动词：to mass-produce（成批生产），to window-dress（布置橱窗），to hand-carry（用

PART I Techniques of EST E-C Translation（科技英语英译汉翻译技巧）

手提）。

从上述例子可以看出，科技英语中大量使用复合名词和复合形容词。复合动词的使用是现代英语中一种比较简练的表现手段。如果不用复合动词来表达，句子会冗长拖沓。但复合动词的使用范围有一定的局限性，不宜任意套用。对这类动词本身，也要区别对待：有临时性词语，时过境迁不再使用；有些富于表现力，能比较长期地保存下来，在翻译过程中，一定要多注意这些词。

(2) 派生词

派生词 Derivation 是由词素（词根、词缀）构成的，词义也是由词素产生的。单词的数量虽然很大，但词素的数量却很有限，词素构成法在英语整个历史发展过程中起着积极的作用，它不仅扩充了英语词汇，而且丰富了语言表现力。英语词缀分前缀（prefix）和后缀（suffix）两种，把它们加在词根的前后，可以构成大量的新词，英语科技翻译者需要掌握足够的词缀和词根，才能识别不断出现的科技英语派生词，为翻译工作打好基础。例如：

词根加前缀的词汇：microreader（缩微阅读器），microminiaturize（使微小化），antinuclear（反对使用核能的），extragalactic（银河系外的），electrochemical（电化学的），macroeconomics（宏观经济学），pseudoscience（伪科学），polygon（多边形），hydrotherapy（水疗法），multimeter（万用表）。

词根加后缀的词汇有：horsemanship（骑术），floriculturist（花卉栽培家），abbreviation（缩写）。

这里要说明的是，无论前缀还是后缀，它们中大部分来源于古英语、拉丁语、希腊语。词根（root）是英语词汇的基本组成部分。绝大多数的自由词根来源于古英语，这类词根在各类词素中是唯一的自由形式；而绝大多数的黏着词根来源于希腊语或拉丁语。由于黏着词根的构词力不强，所以由这种词根构成的词的词义比较专一、稳定，不含感情色彩，没有引申寓意，这些特点使由黏着词根加前后缀构成的词汇，成为科学技术、学术文化所需要的专门术语的好素材。

(3) 混成词

混成词（Blends）是取两个词中在拼写上或读音上比较合适的部分组成一个新词。通常是将两个单词的前部拼接、前后拼接或者将一个单词前部与另一个词拼接构成新的词汇。例如：

camcorder = camera + recorder（摄像机 or video camera）；
comsat = communication + satellite（通信卫星）；
videophone = video + telephone（可视电话）；
bit = binary + digit（二进位数字）；
meld = melt + weld（熔焊）；
digicam = digital camera（数字照相机）；
greentech = green technology（绿色技术）；
informics = information + economics（信息经济学）；
electrochemical = electric + chemical（电气化学的）；
programmatic = program + automatic（能自动编程的）。

混成词在科技英语中比在其他文体中用得多，这是因为混成词便于简略专业术语，便于通过联想理解其含义。

(4) 缩略词

缩略词（Acronym）是将较长的英语单词取其首部或者主干构成与原词同义的短单词，或者将组成词汇短语的各个单词的首字母拼接成一个大写字母的字符串。缩略词在文章索引、前序、摘要、说明书、商标等科技文章中频繁使用。缩略词趋向于任意构词，如某一篇论文的作者可以就仅在该文中使用的术语组成缩略词。缩略词的出现方便了印刷、书写、速记和口语交流等，但是它同时也增加了阅读和理解的困难。所以，为了便于理解，缩略词通常在开始出现时，采用破折号、引号或者括号将它们的原形单词和组合词一并列出，逐渐被人们接受以后，作为注释的后者就消失了。缩略词一般可以分为裁减式缩略词和用首字母组成的缩略词。例如：

裁减式缩略词：maths =（mathematics）数学，lab =（laboratory）实验室，ft =（foot/feet）英尺，cpd（compound）化合物，chute = parachute（降落伞），drome = aerodrome（飞机场）。

用首字母组成的缩略词：FM = frequency modulation（调频），PSI = pounds per square inch（磅/英寸），SCR = silicon controlled rectifier（可控硅整流器），CD = compact disk（激光唱盘），CAD = computer-assisted design（计算机辅助设计），IT = information technology（信息技术），MTV = music television（音乐电视广播），PVC = polyvinyl chloride（聚氯乙烯），ATM = automated teller machine（自动出纳机），AC = alternating current（交流），DC = direct current（直流），CPU = central processor unit（中央处理器单元），MODEM = modulation & demodulator（调制解调器）。

2. Methods of Scientific and Technological Terminology Translation（科技术语的翻译方法）

随着社会的进步和科技的发展，新的发明创造不断涌现，随之就出现了描述这些事物的新术语。在科技英语翻译中，常碰到如何把这类术语译成适当的汉语的问题。通常，有以下约定俗成的方法。

(1) 意译

意译就是对原词所表达的具体事物和概念进行仔细推敲，以准确译出该词的科学概念。这种译法最为普遍，在可能的情况下，科技术语应采用意译法。例如：

holography（全息摄影术），astrionics（宇航电子学），guided missile（导弹），aircraft carrier（航空母舰），videophone（可视电话），walkie-talkie（步话机），power roller（健腹轮），e-mail（电子邮件）等。

一般说来，有三类科技新词语采用此译法：

1) 复合词（Compounding）。例如：

feedback（反馈），skylab（太空实验室），guided missile（导弹），breakdown（击穿），friction factor（摩擦系数），overcharge（超载），waterproof（防水的），moonwalk（月球漫步），walkie-talkie（步话机）等。

2) 多义词（Polysemy）。例如：

bug（窃听器），computer（计算机），monitor（监视器）等。

3) 派生词（Derivation）。例如：

thermocouple（热电偶），voltmeter（电压计），thermometer（温度计），monophase（单相），astrionics（宇航电子学）等。

(2) 音译

音译就是根据英语单词的发音译成读音与原词大致相同的汉字。采用音译的科技术语主要有两类：

1) 计量单位的词。例如：

hertz 赫兹（频率单位），bit 比特（度量信息的单位，二进制位），Ohm 欧姆（电阻单位）lux 勒克司（照明单位），joule 焦耳（功或能的单位），calorie 卡路里（热量单位）。

2) 某些新发明的材料或产品的名称（尤其在初始阶段）。例如：

nylon 尼龙（酰胺纤维），sonar（声呐），vaseline（凡士林）。

一般地说，音译比意译容易，但不如意译能够明确地表达新术语的含义。因此，有些音译词经过一段时间后又被意译词所取代，或者同时使用。例如：

combine（康拜因→联合收割机），laser（莱塞→激光），vitamin（维他命→维生素），penicillin（盘尼西林→青霉素）等。

(3) 形译

科技英语中常用字母的形象来为形状相似的物体定名。翻译这类术语时，一般采用形译法。常用的形译法又分为下列三种情况：

1) 选用能够表达原字母形象的汉语词来译。例如：

T-square（丁字尺），I-column（工字柱），U-bend（马蹄弯头），V-slot（三角形槽），Y-curve（叉形曲线）等。

2) 保留原字母不译，在该字母后加"形"字，这种译法更为普遍。例如：

A-bedplate（A 形底座），D-valve（D 形阀），C-network（C 形网络），M-wing（M 形机翼）等。

3) 保留原字母不译，以字母代表一种概念。例如：

X-ray（X 射线），L-electron［L 层电子（原子核外第二层的电子）］等。

(4) 意音混合译法

意音混合译法是在音译之后加上一个表示类别的词，或者把原词的一部分音译，而另一部分意译。例如：logic（逻辑电路）和 covar（科伐合金/铁镍钴合金）以音译为主，在词首或词尾加上表意的词，而下面几种情况则属于同时使用了音译法和意译法。

1) 由前缀加入计量单位构成复合词，计量单位采用音译。例如：

megavolt［百万伏（特）］，microampere［微安（培）］，kilowatt（千瓦），decibel（分贝）等。

2) 某些复合词意音结合译。例如：

radar-man（雷达手），valve-guide（阀导），motor-cycle（摩托车）等。

3) 有些由人名构成的术语，人名音译，其余部分意译。例如：

Ohm's law（欧姆定律），Curie point（居里点），Morse code（莫尔斯电码），Monel metal（蒙乃尔合金）等。

注意：人们在熟悉了这类术语后往往只记人名的第一音节音译，后加"氏"字译出。例如：

Babbit metal（巴氏合金），Brinell test（布氏试验）等。

(5) 直译

在科技文献中商标、牌号、型号和表示特定意义的字母均可不译，直接使用原文，只译普通名词。例如：

B-52 E bomber（B-52 E 轰炸机），Kubota Mobile Crane Model KM-150（库宝塔 KM-150 型流动式起重机）等。

3. Key Points for Terminology Translation（科技术语翻译要点）

学习了科技术语的翻译方法后，在翻译科技术语时通常还需要注意以下几点：

（1）遵循约定俗成的原则

科技术语的译名应注意规范化。凡约定俗成的译名，不要随意变动。正如鲁迅先生指出的不能把"高尔基"译成"郭尔基"一样。

（2）尽量使用意译法

新术语的译名要能正确表达出事物的真实含义，应尽量采用意译。例如：dustoff（战地救护直升机），该词义与飞机起飞和降落时扬起的尘土有关，dust（尘土）+ off（离开）。

（3）注意专业术语译名的统一性

在同一篇文章或同一本书中专业术语的译名必须前后统一，这特别适用于那些有几个通用译名的专业术语。否则，就可能引起误解。例如：nylon 有"尼龙，耐纶，酰胺纤维"这三个版本的译文，若在同一篇英文文章的译文中出现三者，读者很容易将其看作三个不同的事物，所以同一篇译文应选用其中之一代替 nylon 为好。

4. 翻译时应选择最新版本的词典做参考

当今时代科学技术发展十分迅速，每时每刻都有新的科技事物出现，随着这些科学事物和现象的变化，描述它们的语言也在不断变化着，同一个词语在几年以后会有不同的技术含义，为了准确地翻译科技文献，再现原文内容，建议译者在翻译过程中尽量使用最新版本的技术词典做参考，以免发生误译。

B Introduction to Signals and Systems（信号与系统介绍）

Signals and systems deals with signals, systems, and transforms, from their theoretical mathematical foundations to practical implementation in circuits and computer algorithms, mainly the mathematics and practical issues of signals in continuous and discrete time, linear time-invariant systems, convolution, and Fourier transforms.[1]

1. Definition of Signals and Systems

The notions of signals and systems arise in a wide variety of contexts.

Signals can be used in many ways to describe physical phenomena, such as human speech, voltages over electrical devices, frictional force applied to the automobile, etc. To make it convenient, we represent signals mathematically as functions of one or more independent variables. Especially, when there is only one independent variable, we refer to it as time.

Systems are physically an interconnection of components, devices or subsystems. In information related technology and science, systems can be viewed as a process that input signals are transformed by the system or cause the system to response in some way, resulting in other signals as outputs.[2] For example, an image-enhancement system transforms an input image into an output image that has some desired properties, such as improved contrast.

2. Classifications of signals

1) Periodic vs. aperiodic signals. Periodic signals repeat with some period T, while aperiodic, or non-periodic, signals do not. We can define a periodic function through the following mathematical expression, where t can be any number and T is a positive constant: $x(t) = x(t+T)$

2) Continuous-time vs. discrete-time signals. A continuous-time signal will contain a value for all real numbers along the time axis. In contrast, a discrete-time signal will only have values at equally spaced intervals along the time axis.

3) Analog vs. digital signals. The difference between analog and digital is similar to the difference between continuous-time and discrete-time. However, in this case the difference involves the values of the function. Analog corresponds to a continuous set of possible function values, while digital corresponds to a discrete set of possible function values.

4) Even vs. odd signals. An even signal is any signal $x(t)$ that satisfies $x(t) = x(-t)$. Even signals are symmetric around the vertical axis. An odd signal, on the other hand, is a signal $x(t)$ that satisfies $x(t) = -x(-t)$.

5) Deterministic vs. random signals. Deterministic signals are those that have a fixed behavior and random signals are those that change randomly. For the deterministic signals, each value of them can be determined by a mathematical expression. Thus, we can make accurate assumptions about their past and future behaviors. Unlike deterministic signals, stochastic signals, or random signals cannot be characterized by a simple, well-defined mathematical equation and their future values cannot be predicted. [3]

3. Classifications of Systems

In this section, some of the basic classifications of systems will be briefly introduced.

1) Continuous vs. discrete systems. A system where the input and output signals are continuous is a continuous system, and one where the input and output signals are discrete is a discrete system.

2) Linear vs. nonlinear systems. A linear system is any system that obeys the properties of scaling (homogeneity) and superposition (additivity), while a nonlinear system is any system that does not obey at least one of these. [4]

3) Time invariant vs. time varying systems. A system is said to be time invariant if the behavior and characteristics of the systems are fixed over time. Any system that does not have this property is said to be time varying.

4) Causal vs. non-causal systems. A causal system is one in which the output depends only on current or past inputs, but not future inputs. On contrast, a non-causal system is one in which the output depends on both past and future inputs. All "real-time" systems must be causal, while image processing systems are not necessarily a causal system.

5) Stable vs. unstable systems. A stable system is one in which the output is bounded if the input is also bounded. Similarly, an unstable system is one in which at least one bounded input produces an unbounded output.

New Words and Expressions

implementation　*n.*　实现
Fourier Transform　傅立叶变换
discrete　*adj.*　离散的，不连续的
convolution　*n.*　卷积
variable　*adj.*　易变的，变化无常的；
　n.　[数]变量
component　*n.*　成分，组件，元件
frictional force　摩擦力

image-enhancement system　图像增强系统
periodic　*adj.*　（非）周期的
deterministic　*adj.*　确定性的
stochastic　*adj.*　随机的
scaling　*n.*　缩放（比例）
causal　*adj.*　因果关系的，有原因的
bounded　*adj.*　有界限的

Notes

[1] Signals and systems deals with signals, systems, and transforms, from their theoretical mathematical foundations to practical implementation in circuits and computer algorithms, mainly the mathematics and practical issues of signals in continuous and discrete time, linear time-invariant systems, convolution, and Fourier transforms. 信号与系统从数学理论基础以及电路和计算机算法实践上对信号、系统以及变换进行研究。其中主要研究的是连续和离散时间信号、线性时不变系统、卷积和傅里叶变换的数学算法和应用实践。

[2] In information related technology and science, systems can be viewed as a process that input signals are transformed by the system or cause the system to response in some way, resulting in other signals as outputs. 在信息科学和技术中，系统可以看成一个过程，即系统对输入信号进行变换或者信号使系统以某种方式做出响应，并得到其他信号作为输出。

[3] Unlike deterministic signals, stochastic signals, or random signals cannot be characterized by a simple, well-defined mathematical equation and their future values cannot be predicted. 随机信号不同于确定性信号，它们不能用简单的定义好的数学方程式来表示，并且无法对它们的未来取值进行预测。

[4] A linear system is any system that obeys the properties of scaling (homogeneity) and superposition (additivity), while a nonlinear system is any system that does not obey at least one of these. 满足尺度性（均匀性）和叠加性的系统是线性系统，而二者中任一个条件都不能满足的系统为非线性系统。

C On the Nature of Artificial Intelligence（人工智能的特征）

Artificial intelligence (AI) has grown out of modern computing combined with a plethora of data to interpret and engineering problems to solve. In some sense, it returns to the earlier methods of analyzing data and trying to build predictive models based on empirical data in a "natural" way. In this sense, AI techniques are typically more data-based than dynamics-based. This use of data can make fast, robust, and skillful forecasts possible in domains that might be intractable by a dynamics-based approach. As Einstein once remarked, "So far as the laws of mathematics refer to reality, they are not certain. And so far as they are certain, they do not refer to reality" (Einstein, 1922).

A 19th century scientist would be astonished by the capability of today's computers—solving multi-variable PDEs, running numerical models of the atmosphere, or simulating debris flows, for instance. Yet, that scientist would be equally astonished by the incapabilities of modern computers—their limitations in recognizing a face in a crowd or responding to spoken language, for example. Humans, on the other hand, find that recognizing faces is a cinch but solving multi-variable equations is hard. This is not how it was supposed to be. The earliest computers were billed as machines that could think. Indeed, AI has been defined as enabling machines to perceive, understand, and react to their environments. Although perceptive humanoid robots have been a staple of science fiction for many decades, they are still quite impractical. Rather, successful applications of AI have concentrated on single tasks, such as optimally managing the gates at an airline terminal or successfully classifying cells as benign or cancerous.

AI started out by attempting to build upon Aristotelian ideas of logic. Thus, initial research emphasized induction and semantic queries. The goal was to build a system of logic by which computers could "reason" their way from simple bits of data to complex conclusions. After all, humans do this sort of reasoning effortlessly.[1] However, it slowly became apparent that there is more to human reasoning than just induction. Humans, it turns out, are naturally good at many things in ways that current computer designs may never match.

AI researchers scaled back their overly ambitious initial goal to that of building systems that would complement human users and do tasks that humans found onerous. Computers are good at unflaggingly processing reams of data and performing complex computations, but poor at obtaining a holistic view of systems. Humans, good at higher-level thinking, find it hard to do mind-numbing calculations. AI researchers also approached the problem with a new awareness of the impressive performance of biological systems. Thus, many of the new AI approaches were intentionally modeled on the way human experts thought or behaved or on how underlying biological systems such as the human brain worked.

Rather than building an AI system that would replace a human's multifaceted capabilities, researchers concentrated on building special purpose systems that could do one thing well. Thus, instead of building a system that would conduct a wellness check on a patient, for example, they developed a system that would determine an appropriate drug dosage for a cancer patient. The solutions to such targeted problems were called expert systems, because they encoded the rules that an expert in the field would follow to come to his or her conclusions. Computers proved capable of quickly and objectively determining answers to problems where the methodology was precisely defined. The rules in such expert systems are often in the form of decision trees, where the answer to one question narrows down the possibilities and determines what question is asked next, until all possible conclusions but one (or a few) are eliminated.[2]

One fundamental problem in expert systems is how to represent and apply the domain knowledge of experts. Experts often state their knowledge verbally using imprecise words like "not very hot" and "less water". Such words do not lend themselves well to decision trees since the ambiguity can span multiple branches of the tree at each step. This issue is addressed by another AI technique: fuzzy logic. Fuzzy logic provides a framework for encoding imprecise verbal rules—such as those provided by subject domain experts—and aggregating them to yield a final answer. It also allows partial, ambiguous or uncertain evidence to be maintained and efficiently synthesized in the production of the final result. The most celebrated successes of fuzzy logic have been in Japan, where numerous engineering control systems have been built based on encoding expert knowledge in fuzzy rules that are then combined for prediction or control.

Fuzzy logic can be used to create automated decision support systems that model what a human expert would do under similar circumstances. Because humans are considerably skilled at recognizing patterns and often understand the underlying processes that lead to the data, the verbal rules formulated by human experts are often quite robust, even to unseen data and unanticipated situations. A fuzzy logic system, by piggy-backing on such effective analysis, can possess considerable skill. Because fuzzy logic systems are relatively simple to encode and do not require training datasets, they are also fast to create and implement. Another advantage of these systems, often a deciding factor in many applications, is that the fuzzy rules and their synthesis can be naturally interpreted by a human expert. If a training dataset is available, it can be used to optimize the fuzzy logic algorithm's parameters, though this step is not required. Thus, a fuzzy logic system can provide considerable skill and a human-understandable, tunable system for very little investment.

Fuzzy logic systems often provide good solutions to problems for which reliable expert knowledge is readily available and can be represented with verbal rules or heuristics. While there are many situations where this is the case, there are also many others in which either no experts are available or their knowledge cannot easily be represented with verbal rules. One might gauge the suitability of a fuzzy logic approach by assessing whether different experts tend to agree on data cases; if they don't, it might be necessary to code and evaluate multiple fuzzy logic algorithms to represent the range of solution methodologies, which might not be practical.[3] It may also be difficult to determine whether the verbal rules that experts identify is really all that they use to come to their conclusion. Humans often underestimate the role of intuition or the subconscious knowledge brought to bear on a problem. Moreover, many domains in the environ-mental sciences are exceedingly complex and poorly understood to begin with, so a method capable of automatically recognizing patterns from data may be more appropriate.

Fortunately, another AI method excels at modeling complex, nonlinear systems based on data—the neural network (NN). Like many AI methods, NNs are biologically inspired. The name comes from the fact that they were initially modeled on the way neurons fire, with the accumulated firings of many neurons together determining the brain's response to any particular set of stimuli. The most common architecture used in NNs comprises three layers of neurons—an input layer, a layer of "hidden nodes" and a final output layer. Such an NN can represent any continuous function on a compact domain arbitrarily closely, even a nonlinear one, if it has enough hidden nodes—though choosing the optimal number of hidden nodes for a particular problem may require some effort.[4]

Feed-forward NNs are members of the class of supervised learning machines. In a process called "training", such a learning machine is presented with patterns—sets of inputs and target values, or ideal output corresponding to desired responses to those inputs. The target values may either be provided by an expert in the field, or can be obtained from field surveys, measurements and other information; as such, the target values are often referred to as "ground truth". At the end of training, if all goes well, the learning machine will have created a function that approximately maps the training inputs to the associated targets. Subsequently, when this function is presented with a new set of inputs, it determines a response based on the evidence generalized from the training specimens. Thus, if viewed as an expert system, NNs learn previously unknown relationships or knowledge that experts may not be able to represent with verbal rules. This is because supervised learning machines approximate expert learning behavior, not by approximating the logic that experts use and inexactly describe, but by creating a new mapping to the ground truth associated with the inputs. [5] Specifically, NNs are trained by adjusting their parameters to minimize a cost (or objective) function—a quantity that is usually some function of the difference between the target values and the approximation thereof produced by the network.

Although NNs can represent any continuous function and avoid the problem of depending on expert descriptions by learning data relationships instead, that flexibility comes with a price. First, although the NN learns to approximate the mapping from training samples to target values, the actual function used to represent this approximation is encoded in a large set of connection weights that usually yield no insights. Thus, unlike an expert system, an NN representation is generally not human-understandable, though researchers have found ways to extract approximate rules from an NN in some specific cases. The inability to explain in simple terms the behavior of an NN has led to it being called a "black box". However, it is important to point out that this opaqueness is not specific to NNs but applies to many nonlinear models, which may represent the physical world very well but resist being neatly summarized. [6] The fact is that the human desire to explain relationships in simple terms may be inconsistent with the competing requirement to have the most accurate predictions possible, a trade-off that is not peculiar to AI.

NNs' ability to fit any data places severe requirements on the data necessary for training them. Many NNs have a large number of parameters (weights) that must be estimated during the training phase. The estimates can be highly unreliable if the size of the training data set is not sufficiently large. An abundance of parameters can also lead to over-fitting, which in turn can adversely affect NNs' performance on "new" data (i.e., not included in the training set). In short, properly training an NN requires lots of data.

Another set of biologically-inspired methods are genetic algorithms (GAs). They derive their inspiration from combining the concept of genetic recombination with the theory of evolution and survival of the fittest members of a population. Starting from a random set of candidate parameters, the learning process devises better and better approximations to the optimal parameters. The GA is primarily a search and optimization technique. One can, however, pose nearly any practical problem as one of optimization, including many environmental modeling problems. To configure a problem for GA solution requires that the modeler not only choose the representation methodology, but also the cost function that judges the model's soundness. As mentioned above, training an NN usually involves minimizing some cost function, and that process usually requires differentiating the cost function. By contrast, the learning/training process for a GA does not place any restriction on the differentiability of the cost function, so any measure of performance may be used. The GA is also capable of finding optimal solutions to problems such as those in design. Indeed, genetic algorithms may be used to train either an NN or a fuzzy logic system! Using genetic algorithms to train an NN gives us the ability to use non-differentiable cost functions, while using GAs to train a fuzzy logic system allows us to improve on human-devised rules by optimizing their parameters.[7]

Another method for inferring the relationship between inputs and targets is to automatically build a decision tree based on the training data set. This approach is also among the fastest in terms of training speed: Decision trees can often be trained on substantial data sets in a fraction of the time required by competing techniques. Decision trees, like fuzzy logic systems, also have the advantage of being human-understandable. Unlike fuzzy logic, however, one doesn't need to know the rules beforehand—the rules are learned from training data. Decision trees fell out of favor because the skill realizable with decision trees often lags what is possible using other super vised learning techniques. Recent advances in machine learning—averaging decision trees trained on subsets of the training sets ("bagging") and continually focusing the training on the training data cases that the decision trees get wrong ("boosting") —have made decision trees viable again, but at the cost that the resulting decision trees are no longer human readable.[8] However, aggregate statistics obtained from decision trees are useful in gaining insights into how the decision trees come to their decisions. This is yet another illustration of the aforementioned trade-off between pure performance and transparency.

One of the problems with all of the above data-based methods is that the data on which they are based are always imperfect, corrupted by measurement noise or other artifacts, as are the "ground truth" answers provided in the training data set. Artificial intelligence and statistical methods are closely related in that they both attempt to extract information from noisy data. AI techniques can be utilized to create a practical representation whereas statistical methods can be used to measure how confident we may be that the extracted representation is correct.

New Words and Expressions

a plethora of 大量的
intractable *adj.* 棘手的
cinch *n.* 容易做的事
expert system 专家系统
fuzzy logic 模糊逻辑
dosage *n.* （药品的）用量
synthesize *v.* 人工合成
NN (neural network) 神经网络
decision tree 决策树

approximately *adv.* 近似地
automatically *adv.* 自动地
representation *n.* 表示，陈述
opaqueness *n.* 不透明度
over-fitting 过度拟合
genetic algorithms (GAs) 遗传算法
aggregate *adj.* 总数的，聚合的
transparency *n.* 透明度

PART I Techniques of EST E-C Translation（科技英语英译汉翻译技巧）

Notes

［1］ The goal was to build a system of logic by which computers could "reason" their way from simple bits of data to complex conclusions. 我们的目标是建立一个逻辑系统，在这个系统中，计算机能够从简单的数据点推理得出复杂的结论。

［2］ The rules in such expert systems are often in the form of decision trees, where the answer to one question narrows down the possibilities and determines what question is asked next, until all possible conclusions but one (or a few) are eliminated. 在这种专家系统中，规则通常以决策树的形式出现，缩小问题答案的可能性，且决定下一个问题是什么，直至仅留一个结论，而其他所有可能的结论都被排除在外。

［3］ One might gauge the suitability of a fuzzy logic approach by assessing whether different experts tend to agree on data cases; if they don't, it might be necessary to code and evaluate multiple fuzzy logic algorithms to represent the range of solution methodologies, which might not be practical. 通过评估不同专家的数据情况是否一致，可以获得模糊逻辑方法的适用性；如果不一致，就可能需要对多个模糊逻辑算法进行编码和评估，以表示解决方法的范围，但这种方法可能并不实用。

［4］ Such an NN can represent any continuous function on a compact domain arbitrarily closely, even a nonlinear one, if it has enough hidden nodes—though choosing the optimal number of hidden nodes for a particular problem may require some effort. 虽然对于一个特定的问题选择最佳的隐节点的数目可能需要一些努力，但是如果有足够的隐节点，这种神经网络可以任意近似地表示一个紧致域的任何连续函数，甚至是非线性的。

［5］ This is because supervised learning machines approximate expert learning behavior, not by approximating the logic that experts use and inexactly describe, but by creating a new mapping to the ground truth associated with the inputs. 这是因为监督学习机近似专家学习行为，它不是模拟专家使用的近似而且非精确的逻辑，而是结合输入创造一个真实结果的映射。

［6］ However, it is important to point out that this opaqueness is not specific to NNs but applies to many nonlinear models, which may represent the physical world very well but resist being neatly summarized. 然而，必须指出的是，这种不透明性不只是神经网络特有的，也适用于许多非线性模型，这些模型也许能够很好地代表真实事物，却不能清晰地表示出来。

［7］ Using genetic algorithms to train an NN gives us the ability to use non-differentiable cost functions, while using GAs to train a fuzzy logic system allows us to improve on human-devised rules by optimizing their parameters.
使用遗传算法去训练神经网络使我们能使用非微分损失函数，而使用遗传算法去训练模糊逻辑系统，可以通过优化参数而改进人类设计的（模糊）规则。

［8］ Recent advances in machine learning—averaging decision trees trained on subsets of the training sets ("bagging") and continually focusing the training on the training data cases that the decision trees get wrong ("boosting")—have made decision trees viable again, but at the cost that the resulting decision trees are no longer human readable.

最近机器学习的进步——多个训练子集训练获得的平均决策树（"baggin"算法）和基于决策错误训练集的连续训练（"boosting"算法）——使得决策树再次可行，但是生成决策树的成本却是难以预知的。

Unit 3

A Selecting and Determining the Meaning of a Word（词义的选择和确定）

英汉两种语言都有一词多类和一词多义的现象。一词多类是指一个词往往属于几个词类，具有几个不同的意义；一词多义是指同一个词在同一词类中又往往有几个不同的词义。在英译汉的过程中，在弄清原句结构后，就要善于运用选择和确定原句中关键词词义的技巧，以使所译语句自然流畅，完全符合汉语习惯。选择确定词义通常可以从三方面着手：

1. 根据词性选择词义

有的词不止一个词性，而不同的词性具有不同的含义，因此，翻译时可以通过分析词的词性，从而确定词的具体含义。请看下面的例子：

subject

Example 1：It is known to all that the *subject* of electronics was born from radio.

译文：众所周知，电子学这门<u>学科</u>是从无线电学产生出来的。

Example 2：As a test, the metal was *subjected* to great heat.

译文：该金属曾经<u>受过</u>高温试验。

light

Example 1：All *light* may be traced back to the sun, the sun *lights* our world even after dark.

译文：所有的<u>光</u>都来自于太阳，太阳即使在黑夜后也能<u>照亮</u>我们的世界。

Example 2：The experiences show that any object submerged in water seems to be *lighter*.

译文：经验表明任何物体浸入水以后都似乎变<u>轻</u>了。

Example 3：These facts throw (a) *light* on the subject.

译文：这些事实<u>使</u>问题<u>清楚</u>了。

Example 4：Strike a *light*, please.

译文：请点<u>火</u>。

Like

Example 1：The experiment leads to the conclusion: *like* charges repel; unlike charges attract.

译文：试验得出结论：<u>同性</u>电荷相斥，异性电荷相吸。

Example 2：We can describe a force only by its effects, it cannot be measured directly *like* a length.

译文：我们能够用力的效果来描述一个力，力是不能<u>像</u>长度一样进行直接测量的。

Example 3: He *likes* mathematics more than physics.

译文:他喜欢数学甚于喜欢物理。

Example 4: Wheat, oat, and the *like* are cereals.

译文:小麦、燕麦等诸如此类皆系谷类。

2. 根据专业选择词义

在科技英语中有不少的半技术词汇,这些词汇往往在不同的专业领域会有不同的含义。例如:

cell

生物学:细胞;化学:电解槽;电子学:电池

base

机械学:底座;化学:碱;电子学:基极;数学:(三角形)底边

resistance

resistance to traction:牵引阻力(力学)

resistance to sparking:击穿电压(电子学)

resistance to heat:耐热性(物理学)

resistance to wear:耐磨性(材料学)

solution

Example 1: Many scientists, from their earlier work, have enough knowledge to make good guesses as to the *solution* of a problem they are working on. (在普通英语中)

译文:许多科学家从他们最初工作时起,就完全知道对解决他们正在研究的问题进行适当的猜测。

Example 2: The *solution* has reached such a temperature as not to need to be heated any more. (在化学专业中)

译文:溶液的温度已经足够了,无须再加热。

有时翻译时必须结合专业知识,否则译文就会和原文的含义大相径庭。

Example 1: If a *mouse* is installed in a computer, then the available memory space for user will reduce.

误译:如果让老鼠在计算机里筑窝,那么使用者的记忆空间便会减少。

正确译文:如果计算机安装了鼠标,则用户可利用的内存空间就会减小。

Example 2: Connect the *black pigtail* with the *dog-house*.

误译:把黑色的猪尾巴系在狗窝上。

正确译文:将黑色的引出线接在(高频高压电源的)屏蔽罩上。

3. 根据上下文和搭配关系确定词义

英语词义比较灵活,词的含义范围较宽,词义多变,词义对上下文的依赖性比较大。因而这种灵活的词义,只有结合上下文才能确定下来。例如:

mechanism

Example 1: From this, we can see the *mechanism* of design.

译文:从这点,我们可以看出其设计技巧。

Example 2: Let's see the *mechanism* responsible for typhoons.

译文：让我们分析一下台风形成的<u>过程</u>。

Example 3：An automobile engine is a complex *mechanism*.

译文：汽车发动机是一种复杂的<u>机械装置</u>。

Example 4：We have no *mechanism* for changing the decision.

译文：我们没有<u>办法</u>改变这一决定。

Example 5：At that time, *mechanism* in philosophy became an upstart.

译文：那时，哲学领域的<u>机械主义</u>开始盛行。

Example 6：The *mechanism* of the process, slow and delicate, often escape our attention.

译文：这一<u>发展</u>过程缓慢而细微，我们很难观察到其整个过程。

Example 7：The two model variants of the socialist economic *mechanism* are to be distinguished from the point of view of market function.

译文：从市场功能的角度看，这两种社会主义经济<u>结构</u>的变体是有区别的。

last

Example 1：He is the *last* man to come.

译文：他是<u>最后</u>来的。

Example 2：He is the *last* person for such a job.

译文：他<u>最</u>不配干这个工作。

Example 3：He should be the *last* man to blame.

译文：<u>怎么也不</u>该怪他。

Example 4：This is the *last* place where I expected to meet you.

译文：我怎么<u>也</u>没料到会在这个地方见到你。

除此之外，还可以根据上下文中词的搭配关系来选择和确定词义。例如：

make

Example 1：They are trying to *make* a new type of electric motors to meet the needs of production.

译文：他们正试着<u>生产</u>一种新型的电动机来满足生产的需求。

Example 2：We will *make use of* a number of new sources of energy in the future.

译文：我们将来要<u>利用</u>大量的新能源。

Example 3：Molecules *are made up of* atoms.

译文：分子是由原子<u>组成</u>的。

turn

Example 1：Would you please *turn off* the radio?

译文：请你<u>关掉</u>收音机好吗？

Example 2：She has *turned on* the gas.

译文：她<u>打开</u>了天然气。

Example 3：The company *turns out* a great variety of measuring instruments.

译文：这家公司<u>生产</u>种类繁多的测量仪器。

PART I Techniques of EST E-C Translation（科技英语英译汉翻译技巧）

Exercises

1. Translate the following phrases into Chinese according to the different specialties or the collocations.

order

operational order	order of magnitude
order of a differential equation	order of poles
order of a matrix	give an order for sth.
technical order	

high

high beam	high gear
high brass	high seas
high current	high summer
high explosive	high steel

light

light music	light manners
light loss	light outfit
light car	light work
light heart	light voice
light step	

universal

universal meter	universal use
universal motor	universal class
universal valve	universal travel
universal dividing heading	universal truth
universal constant	universal agent
universal rules	universal peace

2. Translate the following sentences into Chinese by the techniques of selecting and determining the meaning of a word.

power

（1）The fourth <u>power</u> of two is sixteen.

（2）The output <u>power</u> of a machine is always smaller than its input <u>power</u>.

（3）Large quantities of steam are used by modern industry in the generation of <u>power</u>.

（4）Energy is the <u>power</u> to do work.

（5）Friction causes a loss of <u>power</u> in the machine.

（6）China will not be the first to use nuclear weapons although considered one of the nuclear <u>powers</u>.

monitor

（1）Microprocessors <u>monitor</u> tyre wear and brake power on cars.

(2) The patient was connected to a television wave <u>monitor</u>.

develop

(1) To <u>develop</u> the instrument, many experts were invited.

(2) In fact, the seedlings of many tree species will die if fungus does not <u>develop</u> around their roots during their first year of growth.

B Communication Modeling (通信建模)

The first major model for communication came in 1949 by Claude Shannon and Warren Weaver for Bell Laboratories. The original model was designed to mirror the functioning of radio and telephone technologies. Their initial model consisted of three primary parts: sender, channel, and receiver.[1] The sender was the part of a telephone a person spoke into, the channel was the telephone itself, and the receiver was the part of the phone where one could hear the other person. Shannon and Weaver also recognized that often there is static that interferes with one listening to a telephone conversation, which they deemed noise.

In a simple model, often referred to as the transmission model or standard view of communication, information or content (e.g. a message innatural language) is sent in some form (as spoken language) from an emisor/sender/encoder to a destination/receiver/decoder. This common conception of communication simply views communication as a means of sending and receiving information. The strengths of this model are simplicity, generality, and quantifiability. Social scientists Claude Shannon and Warren Weaver structured this model based on the following elements:

1) An information source, which produces a message.

2) A transmitter, which encodes the message into signals.

3) A channel, to which signals are adapted for transmission.

4) A receiver, which decodes (reconstructs) the message from the signal.

5) A destination, where the message arrives.

Shannon and Weaver argued that there were three levels of problems for communication within this theory.

1) The technical problem: how accurately can the message be transmitted?

2) The semantic problem: how precisely is the meaning "conveyed"?

3) The effectiveness problem: how effectively does the received meaning affect behavior?

In 1960, David Berlo expanded on Shannon and Weaver's linear model of communication and created the Sender-Message-Channel-Receiver (SMCR) model of communication. The SMCR model of communication separated the model into clear parts and has been expanded upon by other scholars.

Communication is usually described along a few major dimensions: message (what type of things are communicated), source/emisor/sender/encoder (by whom), form (in which form), channel (through which medium), destination/receiver/target/decoder (to whom), and receiver. Wilbur Schram also indicated that we should also examine the impact that a message has (both desired and undesired) on the target of the message. Between parties, communication includes acts

PART I Techniques of EST E-C Translation（科技英语英译汉翻译技巧）

that confer knowledge and experiences, give advice and commands, and ask questions. These acts may take many forms, in one of the various manners of communication. The form depends on the abilities of the group communicating. Together, communication content and form make messages that are sent towards a destination. The target can be oneself, another person or being, another entity (such as a corporation or group of beings).

Barnlund proposed a transactional model of communication in 2008. The basic premise of the transactional model of communication is that individuals are simultaneously engaging in the sending and receiving of messages.[2]

In a slightly more complex form a sender and a receiver are linkedreciprocally. This second attitude of communication, referred to as the constitutive model or constructionist view, focuses on how an individual communicates as the determining factor of the way the message will be interpreted. Communication is viewed as a conduit; a passage in which information travels from one individual to another and this information becomes separate from the communication itself.[3] A particular instance of communication is called a speech act. The sender's personal filters and the receiver's personal filters may vary depending upon different regional traditions, cultures, or gender; which may alter the intended meaning of message contents. In the presence of "communication noise" on the transmission channel (air, in this case), reception and decoding of content may be faulty, and thus the speech act may not achieve the desired effect.[4] One problem with this encode-transmit-receive-decode model is that the processes of encoding and decoding imply that the sender and receiver each possess something that functions as a code book, and that these two code books are, at the very least, similar if not identical. Although something like code books is implied by the model, they are nowhere represented in the model, which creates many conceptual difficulties.

New Words and Expressions

communication n. 通信
sender n. 发信者，发送人，发报机
channel n. 通道，信道
receiver n. 接收器，接受者，收信机
static n. 静电，静电干扰
destination n. 目的地，终点
quantifiability n. 可定量的性质
transmitter n. 发射机，发射器，发送器
encode v. 编码，译码
decode v. 译码，解码

convey v. 传达
semantic adj. 语义的，语义学的
effectiveness n. 效果，有效性，效率
semiotic adj. 符号学的
transactional model 交易模式，交换理论，互作用模型
conduit n. 导管，管道
co-regulation n. 共同管制，管制，共同规制

Notes

[1] Their initial model consisted of three primary parts: sender, channel, and receiver. 最初的模型主要包括3个部分：发送者、信道和接收者。

[2] The basic premise of the transactional model of communication is that individuals are

simultaneously engaging in the sending and receiving of messages. 通信互作用模型的基本前提是，个体可以同时进行信息的发送和接收。

［3］ Communication is viewed as a conduit; a passage in which information travels from one individual to another and this information becomes separate from the communication itself. 通信看作一个渠道，信息通过它从一个个体传递到另一个个体，并且信息是独立于通信行为而存在的。

［4］ In the presence of "communication noise" on the transmission channel (air, in this case), reception and decoding of content may be faulty, and thus the speech act may not achieve the desired effect. 传输信道（本例指的是大气）中"通信噪声"的存在会使得接收和解码的内容产生差错，因此语言行为达不到预期的效果。

C Artificial Intelligence in Medicine（医疗中的人工智能）

Today, AI is considered a branch of engineering that implements novel concepts and novel solutions to resolve complex challenges. With continued progress in electronic speed, capacity, and software programming, computers might someday be as intelligent as humans. One cannot neglect the important contribution of contemporary cybernetics to the development of AI.

Defined as a trans-disciplinary approach, cybernetics aims for control of any system using technology that explores system regulation, structure and constraints, most notably mechanical, physical, biological, and social. The origin of cybernetics is attributed to Norbert Wiener, who formalized the notion of feedback, with implications for engineering, systems control, computer science, biology, neuroscience, philosophy, and the organization of society.[1] Fields that were most influenced by cybernetics are (if we exclude game theory) systems theory, sociology, psychology (especially neuropsychology and cognitive psychology), and theory of organizations.

Today literature on AI is abundant and unbridled. AI was portrayed as a possible threat to the world economy during the 2015 economic forum held at Davos, where Stephen Hawking even expressed his fear that AI may one day eliminate humanity. We will not discuss here the use of this rapidly developing field in military, security, transport or manufacturing; instead, the focus of this passage is on medicine and health systems.

1. Artificial Intelligence in Medicine: The Virtual Branch

The application of AI in medicine has two main branches: virtual and physical. The virtual component is represented by Machine Learning, (also called Deep Learning) that is represented by mathematical algorithms that improve learning through experience. There are three types of machine learning algorithms: ① unsupervised (ability to find patterns), ② supervised (classification and prediction algorithms based on previous examples), and ③ reinforcement learning (use of sequences of rewards and punishments to form a strategy for operation in a specific problem space). First, AI has boosted and is still boosting discoveries in genetics and molecular medicine by providing machine learning algorithms and knowledge management. An example of successes in medicine is the unsupervised protein—protein interaction algorithms that led to novel therapeutic target discoveries.

The methodology used a combination of adaptive evolutionary algorithms and state-of-the-art clustering methods, named "evolutionary enhanced Markov clustering". It permitted prediction of over 5,000 protein complexes, of which over 70% were enriched by at least one gene ontology function term. Novel computational methodology is also being developed to identify DNA variants such as single nucleotide polymorphisms (SNPs) as predictors of diseases or traits, using novel evolutionary embedded algorithms that are more robust and less prone to over-fitting issues that occur when a model has too many parameters relative to the number of observations.

Today's "systems thinking" about health care not only focuses on the classical interactions between patients and providers but takes into account larger-scale organizations and cycles. Furthermore, the health care system must not be stationary but must learn from its own experiences and strive to implement continuous process improvements. This is a multi-agent system (MAS), where a set of agents situated in a common environment interact with each other. This process involves building or participating in an organization, which uses AI to achieve significant progress.

An example of such a process in medicine is the development of problematically complex ecosystems for treating chronic mental disease. Instead of focusing on health expenditures (in public health systems) or cost recovery (in health management organization), the MAS approach proposes to capture the dynamics of individual patients, including their responses to received medications as well as their behavioral interactions within a larger societal ecosystem.[2] This global care coordination technology allows process mapping, facilitates control, and better supports changes to the system with a demonstrated increase in response to medication, decrease of costs and more efficacious interventions. Its implementation has allowed health systems managers to analyze the dynamics of system performance across changes in social, medical and criminal justice components.

Included in the virtual applications of AI are electronic medical records where specific algorithms are used to identify subjects with a family history of a hereditary disease or an augmented risk of a chronic disease. AI is used to improve organizational performance by enabling individuals to capture, share and apply their collective knowledge to make "optimal decisions in real time". As a consequence, electronic medical records and health care process management are crucial to achieve the desired quality. From current patients' record keeping of variable quality, information needs to be captured in a digital format accessible as individual data as well as in aggregated forms for epidemiological research and planning. Major efforts are required from academia and the information technology industry to achieve desired efficacy and minimize cost.

The current status of medical records is mostly in the form of incommunicable silos of wasted information for the health system and for knowledge acquisition. Laboratories and clinics need to collaborate to accelerate the implementation of electronic health records. Data need to be captured in real-time, and institutions should promote their transformation into intelligible processes. New scientific and clinical findings should be shared through open-source, and aggregated data must be displayed for open-access by physicians and scientists and made automatically available as point-of-care information. Integration and interoperability including ethical, legal and logistical concerns are enormous, particularly with the forthcoming addition of "omicsbased" data. The simplification, readability and clinical utility of data sets should be made evident, and each result must be questioned for its clinical applicability. In the design developed by our group, electronic medical or health records are essential tools for personalized medicine and for early detection and targeted prevention, again with the aim of increasing their clinical value and decreasing health costs.

Further virtual application of AI in medicine is the use of softbots, as psychotherapeutic avatars. Avatars stem from the famous 2009 James Cameron movie which features a hybrid human-alien created to facilitate communication with people from the planet known as Pandora. The use of emotionally sensitive teachable avatars is receiving recognition in medicine. It has been applied to pain control in children with cancer (called "pain body") and it is able to detect early emotional disturbances in youngsters in native American reservations, including suicidal tendency. This approach seems to work better than human interventions. One of the clearest examples is the control of paranoid hallucinations when the subject designs his own avatar representing the persecutor in his mind. The system encourages the subject to engage in discussions with his persecutor who progressively learns to moderate such destructive behavior. Initial successes with this technology have been demonstrated by achievement of a lower level of hallucinations and vocal threats. Perhaps the most useful function will be in care of the elderly, where the frequency, reassuring nature, and kindness of what is said are all important elements of improved communication. Avatars have been also applied to home care, and for biological and physical monitoring with 3D vision.

2. Artificial Intelligence in Medicine: The Physical Branch

The second form of application of AI in medicine includes physical objects, medical devices and increasingly sophisticated robots taking part in the delivery of care (carebots). Perhaps the most promising approach is the use of robots as helpers; for example, a robot companion for the aging population with cognitive decline or limited mobility. Japanese carebots are the most advanced forms of this technology. Robots are used in surgery as assistant-surgeons or even as solo performers. One of the most impressive examples of the utility of robots is their ability to communicate with and teach autistic children. Here, and in many other situations that might benefit from robotic intervention, important ethical considerations will have to be resolved before it will become possible to use AI-robots routinely in today's medical environment. Apart from ethical issues, a major challenge in this new dimension of medical care is the clear need for standardized, comparative evaluation of the effect of robotic systems on health indicators, and measures of changes in psychological and physical status, side effects, and outcomes.[3]

3. Use of Robots to Monitor Effectiveness of Treatment

Robots can also be useful in the evaluation of changes in human performance in such situations as rehabilitation. Another area where AI might be helpfully employed is for monitoring the guided delivery of drugs to target organs, tissues or tumors. For example, it is encouraging to learn of the recent development of nanorobots designed to overcome delivery problems that arise when difficulty of diffusion of the therapeutic agent into a site of interest is encountered.[4] This problem occurs when the therapist is attempting to target the core of a tumor which tends to be less vascularized, anoxic, but most proliferatively active. To overcome limitations of mechanical or radioactive robotics, researchers have attempted to harness a natural agent with desired properties as a replacement of "intelligent" nanoparticles alone. For this purpose, they are studying a special type of marine coli, called Magnetococcus marinus which travels spontaneously to low oxygenated zones. Initial guidance is provided by an external magnetic source and then inherent properties of nanorobots are put into play. These nanorobots can be covalently bound with nanoliposomes bearing therapeutic properties. Early data have disclosed a significant increase in the gradient of desired drug into the hypoxic zones.

Most of these novel applications of AI in medicine need further research, particularly in areas of human-computer interactions. Moshimo Mori introduced in 1970 the notion of uncanny valley in which an important theme is the human-robot interaction (HRI) field. In these studies humanoid robots were evaluated for their apparent humanity, eeriness and attractiveness as factors making perception of robots either acceptable, feared or rejected.

AI for personal use is going to stay with us much as genetics will continue to provide personal services. It is therefore important to consider how AI will also serve the development of our health care systems. Takashi Kido proposed MyFinder as a personalized community computing to resolve challenges of personalized genome services, acting jointly with AI and shaping the personalized and participative health care of the future. The goal of this platform is to provide personal genome environment interaction in both directions: impact of genes on diseases, health and drug responses, and impact of our environment, behavioral and wellness on our gene activities. The World Economic 2016 Forum named open AI ecosystem as one of the 10 most important emerging technologies. With the unprecedented amount of data available, combined with advances in natural language processing and social awareness algorithms, applications of AI will become increasingly more useful to consumers. This is particularly true in medicine and healthcare where there are many data to be utilized from patient medical records and lately also from information obtained by wearable health sensors. This huge volume of data should be analyzed in detail, not only to provide patients who want suggestions about lifestyle, but also to generate information aimed at improving healthcare design, based on the needs and habits of patients. It is important to tear down the prejudices and fears regarding AI and understand how it could be beneficial and how we can cope with its perceived or real drawbacks. [5] The biggest apprehension we have is that AI will become so sophisticated that it will surpass human brain capabilities and eventually will take control over our lives. However, if we succeed in creating ethical standards, developing measures of success and effectiveness, making it available to the mainstream, and not only to the Ivy League medical institutions, by making AI tools open-source and user-friendly and of proven clinical utility, then societal benefits will accrue from the use of AI. [6]

New Words and Expressions

a branch of ……的一个分支
contemporary cybernetics 当代控制论
trans-disciplinary *adj.* 跨学科的
game theory *n.* 博弈论

neuropsychology *adj.* 神经学的
unbridled *adj.* 放肆的，无拘束的
portray *v.* 描绘，扮演
unsupervised *adj.* 无人监督的，无人管理的
supervised *adj.* 有监督的
reinforcement *n.* 加强，加固
molecular *adj.* 分子的，由分子组成的
evolutionary enhanced Markov clustering 进化增强马尔可夫聚类
SNPs (single nucleotide polymorphisms) 单核苷酸多态性
embedded *adj.* 嵌入式的
stationary *adj.* 固定的
MAS (multi-agent system) 多智能体系统，多代理系统

ecosystems *n.* 生态系统
intervention *n.* 干预，干涉
epidemiological *adj.* 流行病学的，传染病学的
incommunicable *adj.* 不可言喻的
clinical applicability 临床适用性
psychotherapeutic avatars 心理分析的化身
persecutor *n.* 迫害者
hallucination *n.* 幻觉
autistic *adj.* 患自闭症的；患孤独症的
proliferatively *adv.* 增殖地
put into play 投入使用
eeriness *n.* 怪诞，阴森
genome *n.* 基因组
prejudice *n.* 偏见

Notes

[1] The origin of cybernetics is attributed to Norbert Wiener, who formalized the notion of feedback, with implications for engineering, systems control, computer science, biology, neuroscience, philosophy, and the organization of society. 控制论的起源归功于诺伯特·维纳，他将反馈概念规范化，并涉及了工程、系统控制、计算机科学、生物学、神经科学、哲学和社会组织的领域。

[2] Instead of focusing on health expenditures (in public health systems) or cost recovery (in health management organization), the MAS approach proposes to capture the dynamics of individual patients, including their responses to received medications as well as their behavioral interactions within a larger societal ecosystem. 不是关注于医疗保健支出（公共卫生系统）或成本回收（健康管理组织），MAS 方法提出了捕获个体患者的动态，包括他们对受到药物治疗的反应，以及他们在一个更大的社会生态系统的行为交互。

[3] Apart from ethical issues, a major challenge in this new dimension of medical care is the clear need for standardized, comparative evaluation of the effect of robotic systems on health indicators, and measures of changes in psychological and physical status, side effects, and outcomes. 除了伦理问题外，医疗保健这一新领域的主要挑战是明显需要对机器人系统在健康指标、心理和生理变化测度、副作用和疗效方面的影响进行标准化、比较性的评估。

[4] For example, it is encouraging to learn of the recent development of nanorobots designed to overcome delivery problems that arise when difficulty of diffusion of the therapeutic agent into a site of interest is encountered. 例如，最近令人振奋的研究进展是可用于克服药物输送问题的纳米机器人，这种问题通常发生在治疗药物注入病灶区扩散发生困难的时候。

[5] It is important to tear down the prejudices and fears regarding AI and understand how it could be beneficial and how we can cope with its perceived or real drawbacks. 重要在于打破对人工智能的偏见和恐惧，并懂得如何使其有益于人类，以及我们如何应对它的感知能力和实际缺点。

[6] However, if we succeed in creating ethical standards, developing measures of success and effectiveness, making it available to the mainstream, and not only to the Ivy League medical institutions, by making AI tools open-source and user-friendly and of proven clinical utility, then societal benefits will accrue from the use of AI. 然而，如果我们成功地建立了道德标准，开发成功和有效的措施，通过人工智能工具的开源和用户友好界面，以及临床验证使它不仅在常春藤医疗机构中可以成为主流，而且社会效益也将通过使用人工智能而增加。

Unit 4

A The Extension of the Meaning of a Word or a Phrase（词义的引申）

1. 词义引申的方法

在科技英语翻译过程中，有时会遇到在词典上找不到与原文相符的对应词，或虽然有对应词，但不符合汉语的表达习惯，或出于修辞等原因，词的字面意义与表达意义不一致等情况。遇到上述情况时，就要使用引申这一翻译技巧。所谓引申，就是从原词的基本意义出发，将其改变成一种适于表达原文精神实质的新词义。从翻译的角度看，引申可分为两个方面：一是英语本身对所用词的本义所做的调整与变动，这对于译者来说，主要是理解问题；二是原文基本词义确凿，只是译文表达时需做适当变动。通过下面的例句可以仔细体会如何引申词义。

（1）名词词义的引申

Example 1：The study of neural network is one of the *last frontiers* of artificial intelligence.

译文：对神经网络的研究是人工智能的最新领域之一。

Example 2：All the *wit and learning* in this field are to be present at the symposium.

译文：所有的这一领域的学者都将出席这个科学讨论会。

Example 3：The contributors in component technology are the semiconductor components.

译文：元件技术中起主要作用的是半导体元件。

Example 4：IPC (Industrial Personal Computer) took over an immense range of tasks from worker's *muscles and brains*.

译文：工控机取代了工人大量的体力劳动和脑力劳动。

Example 5：The foresight and *coverage* shown by the inventor of this apparatus are impressing.

译文：这种装置的发明者所表现的远见和渊博学识给人很深的印象。

Example 6：An earthquake occurs when the rocks are strained to *failure*.

译文：当岩石受拉发生断裂时地震就发生了。

Example 7：The *existence* of a differential equation does not imply the *existence* of any solution of equation.

译文：微分方程的存在不意味着方程就有解。

Example 8：Colors can give more *force* to the form of the product.

译文：色彩能给产品的外观增添魅力。

Example 9：The *shortest distance* between raw materials and finished parts is casting.

译文：将原料变成成品的捷径是铸造。

（2）动词词义的引申

work

Example 1：This method *works* well.

译文：这种方法效果良好。

Example 2：My watch doesn't *work*.

译文：我的表停了。

Example 3：The machine *works* smoothly.

译文：这台机器运转正常。

Example 4：Vibration has *worked* some connection loose.

译文：震动使得一些连接元件松动了。

Example 5：The instrument is not *working* well.

译文：这台仪表失灵了。

Example 6：The movement of the spring is made to *work* a pointer on a dial so weight is recorded.

译文：弹簧伸缩带动刻度盘指针，从而记录下重量。

Example 7：*Working* with numerals, a computer is similar in many aspects to an automatic language translator.

译文：用数字进行运算的计算机在很多方面和自动语言翻译机很相似。

Example 8：The interference *worked* much unstability in the system.

译文：干扰给系统造成了很大的不稳定性。

（3）形容词词义的引申

Example 1：Our bodies tend to become *upset* in weightless conditions.

译文：在失重情况下，人体往往失去了平衡。

Example 2：The atomic clock is *accurate* to one hundredth of a second within a year.

译文：原子钟每年误差不超过百分之一秒。

Example 3：He rose to become the *leading* mathematician of his age.

译文：他成长为当时第一流的数学家。

（4）副词词义的引申

Example 1：The thicker the wire, the more *freely* it will carry current.

译文：导线越粗，导电就越容易。

Example 2: *Here* we have a potential difference, and yet not current.
译文：<u>在这种情况下</u>，有电位差而没有电流。

2. 词义引申要点

词义引申是英译汉时常用的技巧之一。在词义引申时还需要注意以下几个问题：

（1）需要译者具有较强的逻辑思维和推理能力

科技文章具有较强的科学性、逻辑性，词与词之间，段与段之间总是互相依存、相互制约的。一般说来，需要引申的英文多见于单个的词、词组或短语，引申的背景又多是现象、事理或与主题相关的概念。因此，在进行引申处理时，要特别注意从原词的基本意义出发，根据上下文的关系进行逻辑推理，准确把握该词、词组等在原作者笔下的实际意义。引申翻译的词义，要求概念明确清楚、逻辑关系清晰突出，文字简洁明了，符合技术术语表达的习惯，体现科技英语翻译的科学、准确、严谨的特征。

（2）要根据词的联立关系正确理解词义

所谓"词的联立关系"是指词在行文中的搭配、组合关系。一个孤立的英语单词，其词义总是游移不定的，具有该词可能具有的一切词义。但当词处于特定的联立关系时，它的词义受到毗邻词的制约而变得稳定化、明朗化了。因此，在进行引申处理时，要注意根据词的联立关系，理解基本词义，决不能孤立、片面、静止地去理解一个词的词义，拿一个词义到处套用，或者超出基本词义所允许的范围而随意发挥。例如，heavy 的基本词义是"重"，heavy crop 可以引申为"大丰收"，heavy current 可以引申为"强电流"，heavy traffic 可以引申为"交通拥挤"，但不能把 heavy industry（重工业）引申为"大规模的工业"。

（3）要符合汉语的表达习惯

翻译的目的是给不懂外文的读者提供一个译本，使其读后能得到如同读原著一样的感受。为此，对译文的基本要求，除了忠实地表达原文的精神实质，再现原文的风格外，还要求语言必须通顺流畅，即译文必须是地道的中文，是中国人最熟悉的语言风格，最符合中文表达规范的语言形式。例如，在翻译法国雷诺汽车的英文版使用说明书时，有两句原文："Drive enthusiastically is expensive in fuel. Drive with a 'light foot'."用中文按字面直译，总觉得说不清楚，感到别扭。而经过反复琢磨后，将其原意引申译成"猛跑狂开耗油多，脚下留情才合算"。译文用的是汉语对称结构，读之朗朗上口，好懂易记。又如："I'm too old a dog to learn new tricks."（我们上了年纪，学不会新道道儿了）。原文中的形象和喻义的结合关系对中国人来说很陌生，如直译过来，中国人不易接受，而若按汉语的习惯翻译，将其原意引申，就能使中国读者一目了然。引申是语言的普遍现象。不同的文体引申的范围、内容、方式不尽相同。要想将引申运用得自如、恰到好处，除了必须掌握的技巧之外，译者自身的语言基本功、文化修养、知识面等都是很重要的。

Exercises

Translate the following sentences into Chinese by the extension techniques of the meaning of a word or a phrase. Make sure to translate the underlined words or phrases properly.

（1）Atoms are much too small to be seen even through the <u>most powerful</u> microscope.

（2）The study of the brain is one of the <u>last frontiers</u> of human knowledge and of much more immediate importance than understanding the infinity of space or the mystery of the atom.

(3) We see that the surface is covered with tiny "hills and valleys".
(4) The beauty of laser is that it can do machining without ever physically touching the material.
(5) The major problem in fabrication is the control of contamination and foreign materials.
(6) Every life has its roses and thorns.
(7) The energy of the sun comes to the earth mainly as light and heat.
(8) A network of highway was built from coast to coast.
(9) He found that mercury column measured the predicted height.
(10) Public opinion demands that something should be done to end the strike.
(11) The character of these people is a mixture of the tiger and the ape.
(12) There are threesteps which must be taken before we graduate from the integrated circuit technology.
(13) The expense of such an instrument has discouraged its use.
(14) A creative person will almost never follow a set pattern in developing an idea. To do so would tend to structure his thinking and might limit possible solutions.
(15) In a DC circuit, current and voltage measures the amount of power.

B Typical DSP Applications（数字信号处理的典型应用）

1. Telecommunications

Telecommunications is about transferring information from one location to another. This includes many forms of information-telephone conversations, television signals, computer files, and other types of data. To transfer the information, you need a channel between the two locations. This may be a wire pair, radio signal, optical fiber, etc. Telecommunications companies receive payment for transferring their customer's information, while they must pay to establish and maintain the channel. The financial bottom line is simple—the more information they can pass through a single channel, the more money they make. DSP (Digital Signal Processor) has revolutionized the telecommunications industry in many areas—signaling tone generation and detection, frequency band shifting, filtering to remove power line hum, etc.[1] One specific example from the telephone network will be discussed here—multiplexing.

Multiplexing

There are approximately one billion telephones in the world. At the press of a few buttons, switching networks allow any one of these to be connected to any other in only a few seconds. The immensity of this task is mind boggling! Until the 1960s, a connection between two telephones required passing the analog voice signals through mechanical switches and amplifiers. One connection required one pair of wires. In comparison, DSP converts audio signals into a stream of serial digital data. Since bits can be easily intertwined and later separated, many telephone conversations can be transmitted on a single channel. For example, a telephone standard known as the T-carrier system can simultaneously transmit 24 voice signals. Each voice signal is sampled 8,000 times per second using an 8 bit companded (logarithmic compressed) analog-to-digital conversion. This results in each voice signal being represented as 64,000 bits/sec, and all 24 channels being contained in 1.544 megabits/sec. This signal can be transmitted about 6,000 feet using ordinary telephone lines of 22 gauge copper wire, a typical interconnection distance. The financial advantage of digital transmission is enormous. Wire and analog switches are expensive; digital logic gates are cheap.

2. Audio Processing

The two principal human senses are vision and hearing. Correspondingly, much of DSP is related to image and audio processing.

(1) Music

The path leading from the musician's microphone to the audiophile's speaker is remarkably long. Digital data representation is important to prevent the degradation commonly associated with analog storage and manipulation. This is very familiar to anyone who has compared the musical quality of cassette tapes with compact disks. In a typical scenario, a musical piece is recorded in a sound studio on multiple channels or tracks. In some cases, this even involves recording individual instruments and singers separately. This is done to give the sound engineer greater flexibility in creating the final product. The complex process of combining the individual tracks into a final product is called mix down. DSP can provide several important functions during mix down, including: filtering, signal addition and subtraction, signal edition, etc. One of the most interesting DSP applications in music preparation is artificial reverberation. If the individual channels are simply added together, the resulting piece sounds frail and diluted, much as if the musicians were playing outdoors.[2] This is because listeners are greatly influenced by the echo or reverberation content of the music, which is usually minimized in the sound studio. DSP allows artificial echoes and reverberation to be added during mix down to simulate various ideal listening environments. Echoes with delays of a few hundred milliseconds give the impression of cathedral like locations. Adding echoes with delays of 10 − 20 milliseconds provide the perception of more modest size listening rooms.

(2) Speech Generation

Speech generation and recognition are used to communicate between humans and machines. Rather than using your hands and eyes, you use your mouth and ears. This is very convenient when your hands and eyes should be doing something else, such as: driving a car, performing surgery, or (unfortunately) firing your weapons at the enemy. Two approaches are used for computer generated speech: digital recording and vocal tract simulation. In digital recording, the voice of a human speaker is digitized and stored, usually in a compressed form. During playback, the stored data are uncompressed and converted back into an analog signal. An entire hour of recorded speech requires only about three megabytes of storage, well within the capabilities of even small computer systems. This is the most common method of digital speech generation used today. Vocal tract simulators are more complicated, trying to mimic the physical mechanisms by which humans create speech. The humanvocal tract is an acoustic cavity with resonate frequencies determined by the size and shape of the chambers. Sound originates in the vocal tract in one of two basic ways, called voiced and fricative sounds. With voiced sounds, vocal cord vibration produces near periodic pulses of air into the vocal cavities.[3] In comparison, fricative sounds originate from the noisy air turbulence at narrow constrictions, such as the teeth and lips. Vocal tract simulators operate by generating digital signals that resemble these two types of excitation. The characteristics of the resonate chamber are simulated by passing the excitation signal through a digital filter with similar resonances. This approach was used in one of the very early DSP success stories, the Speak & Spell, a widely sold electronic learning aid for children.

3. Image Processing

Images are signals with special characteristics. First, they are a measure of a parameter over space (distance), while most signals are a measure of a parameter over time. Second, they contain a great deal of information. For example, more than 10 megabytes can be required to store one second of television video. This is more than a thousand times greater than for a similar length voice signal. Third, the final judge of quality is often a subjective human evaluation, rather than an objective criteria. These special characteristics have made image processing a distinct subgroup within DSP.

New Words and Expressions

acoustic adj. 声学的，听觉的	studio n. 工作室，演播室，画室，电影制片厂
immensity n. 巨大，无限，广大	perception n. 感觉，知觉，洞察力，看法
intertwine v. 交错	recognition n. 识别
degradation n. 退化，降格，降级，堕落	surgery n. 外科，外科手术
mix down 缩混，混合	vocal tract 声道
frail adj. 虚弱的，脆弱的	voiced adj. 浊音的，有声的
revolutionize v. 革命化	fricative n. 摩擦音 adj. 摩擦音的，由摩擦产生的
mind-boggling adj. 难以置信的	subjective adj. 主观的，个人的
audiophile n. 唱片爱好者，爱玩高级音响的人	objective adj. 客观的，目标的
scenario n. 方案，情节，剧本	resonate v. 共鸣，共振
diluted adj. 无力的	

Notes

[1] DSP has revolutionized the telecommunications industry in many areas—signaling tone generation and detection, frequency band shifting, filtering to remove power line hum, etc. 数字信号处理革新了电信工业的很多领域——信号的发生和检测，频带飘移，电力线干扰的滤除等。

[2] If the individual channels are simply added together, the resulting piece sounds frail and diluted, much as if the musicians were playing outdoors. 如果单个通道简单地组合在一起，获得的声音片段既支离破碎又模糊，很像音乐家在室外弹奏的音乐。

[3] With voiced sounds, vocal cord vibration produces near periodic pulses of air into the vocal cavities. 发浊音时，声带震动产生近似于周期脉冲气流至声腔内。

C Fundamentals of IoT(物联网基础)

1. Design Pattern for IoT

The goal of IoT (Internet of things) is to join different objects/things over the networks. As a key technology in integrating varied systems or devices, the service-oriented architecture (SOA) is a design pattern that can be applied to support IoT. The architecture has been effectively applied in research areas within the scopes of cloud computing, WSNs (wireless sensor networks), and vehicular network. Several concepts are already offered to produce multi-layer service-oriented architectures for IoT based on the selected technology, business technical requirements. As an example, the International Telecommunication Union (ITU) recommends that IoT design involves five diverse layers: sensing, accessing, networking, middleware, and application layers. The work of suggests to split the IoT system architecture into three main layers: perception layer, network layer, and service layer (sometimes labeled as the application layer). The authors in created a three-layered architectural model for IoT that contains the application layer, the network layer, and the sensing layer. The authors in designed and created an IoT application infrastructure that covers the physical layer, transport layer, middleware layer, and applications layer.

The scheme in an architectural manner of IoT is concerned with architecture styles, web services and applications, smart objects, networking and communication, business models and corresponding process, cooperative, security, data processing, etc.[1] From the technology viewpoint, the design of an IoT architecture must consider extensibility, scalability, modularity, and interoperability among mixed devices. As things might move or need real-time interaction with their surroundings, an adaptive architecture is required to assist devices dynamically to interact with other objects. The decentralized and heterogeneous nature of IoT needs that the architecture delivers IoT effectual event-driven competence. Thus, SOA (service-oriented architecture) is being seen as a good method to attain inter-operability among heterogeneous devices in a multitude of way.

(1) Sensing Layer

IoT is to be seen as a universal physical inner-connected net, where objects can be linked and controlled remotely. As we observe more devices equipped with RFID (radio frequenly identification devices) or brainy sensors, linking things becomes much more trivial. In the sensing layer, the wireless smart systems with tags or sensors are currently able to sense and exchange information among different devices fully automatically. This technology improves expressively the capability of IoT to sense and identify things or environment. In some business sectors, intelligent service deployment systems and a universal unique identifier (UUID) are allocated to each service or device that may be wanted. A device with UUID can be simply recognized and retrieved. Therefore, UUIDs are critical for effective services deployment in a giant network like IoT.

(2) Networking Layer

The part of networking layer is to link all things with each other and permit things to share the information with other associated things. Furthermore, the networking layer is talented of combining information from current IT infrastructures (i.e., transportation systems, business systems, power grids, healthcare systems, ICT systems, etc.). In SOA-IoT, services given by things are classically deployed in a mixed network and all related things are carried into the service Internet. This procedure might include QoS (quality of service) management and control according to the necessities of users/applications. Instead, it is important for a dynamically altering network to automatically determine and plot things in a network. Things must be automatically allocated with roles to organize, manage, and plan the behaviors of things and be talented to shift to any other roles at any time as desired. These abilities permit devices to be able to collaboratively complete tasks. When designing the networking layer in IoT, designers need to address topics such as network management technologies for heterogonous networks (such as fixed, wireless, mobile, etc.), energy efficiency in networks, QoS requirements, service discovery and retrieval, privacy and security, and data and signal processing. [2]

(3) Service Layer

The service layer is depended on the middleware technology that offers methods to flawlessly assimilate services and applications in IoT. The middleware technology offers the IoT with a cost efficient stand, where the hardware and software stages can be reapplied. A main job in the service layer includes the service specifications for middleware, which are being created by numerous organizations. A well-designed service layer is capable to classify common application necessities and deliver APIs (application program interface) and protocols to sustenance required services, user needs, and applications. [3] This layer also processes all service-oriented difficulties, i.e., information exchange and storage, data management, search engines, and communication. This layer includes ① service discovery: finding objects that can give the needed services and information in a well-organized way; ② service composition: enabling the communication among connected things; ③ trustworthiness management: pointing at determining trust and reputation functions that can assess and Internet of things in real-life—a great understanding apply the information provided by other services to create a reliable system; ④ service APIs: supporting the relations between services required in IoT.

(4) Interface Layer

Within IoT, a great number of devices involved are created by dissimilar manufacturers/vendors and they do not permanently fulfill the exact standards/protocols. This concludes in many interaction problems with exchange of information, communication, and cooperative event processing among different things. Also, the continuous increase of things contributing in an IoT makes it more difficult to dynamically connect, communicate, disconnect, and operate. There is also a need for an interface layer to make the management and interconnection of things easier. An interface profile (IFP) can be understood as a division of service standards that support interaction with applications deployed on the network. A decent interface profile is related to the implementation of universal plug and play (UPnP), which describes a protocol for enabling interaction with services provided by various things. The interface profiles are applied to define the specifications between applications and services. The services on the service layer run directly on limited network infrastructures to successfully find new services for an application, as they connect to the network. [4] Lately, an integration architecture has been suggested to successfully interact among applications and services. Traditionally, the service layer provides universal API for applications. However, the current research results on SOA-IoT reported that service provisioning process (SPP) can also efficiently offer interaction between services and applications.

2. Technologies

(1) Communication Technology

IoT may include many electronic devices, mobile devices, and industrial equipment. Dissimilar things have diverse communication, networking, data processing, data storage capacities, and transmission power. For example, numerous smart phones today have strong communication, networking, data processing, and data storage volumes. Associated to smart phones, heart rate monitor watches only have limited communication and computation abilities. All these objects can be linked by networking and communication technologies.

IoT includes several assorted networks such as WSNs, wireless mesh networks, and WLAN. Those networks assist objects in IoT to exchange information. A gateway has the skill to enable the communication or interaction of numerous devices over the Internet. The gateway can also influence its network information by performing optimization algorithms locally. Thus, a gateway can be applied to grip many multifaceted features involved in communication on the network.

Dissimilar objects may have changing QoS requirements, for example performance, energy efficiency, and security. Let us say, a lot of devices depend on batteries and thus reducing energy usage for these devices is a top concern. In distinction, devices with power supply connection frequently do not set energy saving as a top precedence. IoT could also importantly benefit by leveraging current Internet protocols such as IPv6, since this will make it possible to directly address any number of things required through the Internet.[5] The leading communication protocols and standards are NFC, RFID, IEEE 802.15.4 (ZigBee), IEEE 802.11 (WLAN), IEEE 802.15.1 (Bluetooth), multihop wireless sensor/mesh networks, IETF low power wireless personal area networks (6LoWPAN), traditional IP technologies such as IP and IPv6, and machine to machine (M2M).

(2) Tracking/Identification Technology

The tracking and identification technologies within IoT are RFID systems, barcode, and intelligent sensors. A usual RFID system is a combination of an RFID reader and an RFID tag. The RFID system is gradually being applied, such as logistics, supply chain management, and healthcare service monitoring, due to its skill to identify, trace, and track devices and physical objects. Other pros of the RFID system include offering detailed real-time information about the involved devices, dropping labor cost, shortening business process, enlarging the accuracy of inventory information, and improving business efficiency.[6] Modern development of the RFID technology focuses on the following aspects ① active RFID systems with spread-spectrum transmission; and ② technology of managing RFID applications.

Still, there is a large space for the development of the RFID-based applications. To further promote the RFID technology, RFID can be combined with WSNs to better track and trace things in real-time.

(3) Service Management

Service management is the way to implement and manage quality of IoT services that meet the requirements of applications or users. The SOA can be applied to summarize services by hiding the implementation details of services such as protocols applied. Here it is possible to decouple between components in a system and therefore hide the heterogeneity from end users. An SOA-IoT lets application to apply heterogeneous objects as compatible services. Instead, the lively nature of IoT applications needs IoT to deliver consistent and reliable services. An actual SOA can diminish the influence caused by device moves or battery failure.

A service is a group of data and associated performances to achieve a particular function or feature of a device or portions of a device. A service might locate other main or subordinate services and/or a set of characteristics that make up the service. The services can be characterized into two types: main service and secondary service. The first mentioned means services that represent the primary functionalities, which can be seen as the elementary service component and can be included by another service.[7] The last mentioned secondary service can offer supplementary functionalities to the primary service or other secondary services. A service can consist up to many features, which describes service data structures, descriptors, permission, and other attributes of a service.

(4) Network

There are not so many cross-layer protocols for wireless networks, i.e., wireless sensor and actuator networks (WSANs) or ad hoc networks (AHNs). Still, they need vision beforehand such that they can be applied to the IoT. The purpose is that IoT often have varied communication and computation capabilities, and varying QoS requirements. In difference, nodes in WSNs classically have similar necessities for hardware and network communication. In addition, the IoT network applies the Internet to support information exchange and data communication. In difference, WSNs and AHNs do not have to include the Internet for communication.

New Words and Expressions

IoT (Internet of things) 物联网
integrate v. 整合，积分
SOA (service-oriented architecture) 面向服务的体系结构
WSNs (wireless sensor networks) 无线传感器网络
ITU (International Telecommunication Union) 国际电信联盟
infrastructure n. 基础设施
scalability n. 可扩展性
modularity n. 模块化
interoperability n. 互操作性
dynamically adv. 动态地，不断变化地

decentralized adj. 分散的
heterogeneous adj. 各种各样的
RFID (radio frequency identification devices) 无线射频识别
trivial adj. 琐碎的
UUID (universal unique identifier) 通用唯一标志符
ICT (information & communication technology) 信息与通用技术
QoS (quality of service) 服务质量
allocate v. 分配
flawlessly adv. 完美地
API (application program interface) 应用程

| 序接口 | 路通信协定第六版 |

sustenance *n.* 营养
dissimilar *adj.* 不同的，不相似的
IFP（interface profile） 界面特征
deploy *v.* 部署
UPnP（universal plug and play） 通用即插即用
specification *n.* 规格，说明书
provision *n.* 供应
WLAN（wireless LAN） 无线局域网
gateway *n.* 网关
optimization *n.* 最优化
multifaceted *adj.* 多方面的
IPv6（Internet Protocol Version 6） 网际网

NFC（near field communication） 近距离无线通信技术
IEEE（Institute of Electrical and Electronic Engineers） 电气和电子工程师学会
M2M（machine to machine） 机对机
barcode *n.* 条码
spread-spectrum 扩频
decouple *v.* 解耦
heterogeneity *n.* 异质性
attribute *n.* 属性
WSANs（wireless sensor and actuator networks） 无线传感器与执行器网络
AHNs（ad hoc networks） 无线自组网

Notes

［1］The scheme in an architectural manner of IoT is concerned with architecture styles, web services and applications, smart objects, networking and communication, business models and corresponding process, cooperative, security, data processing, etc. 物联网的构造方案涉及构造风格、网络服务和应用程序、智能对象、网络和通信应用、商业模式，以及相应的过程、合作、安全、数据处理等。

［2］When designing the networking layer in IoT, designers need to address topics such as network management technologies for heterogonous networks (such as fixed, wireless, mobile, etc.), energy efficiency in networks, QoS requirements, service discovery and retrieval, privacy and security, and data and signal processing. 在设计物联网的网络层时，设计师需要解决的问题很多，如异构网络（固定、无线、移动等）的网络管理技术、网络运行的效率、服务质量要求、服务发现和检索、隐私和安全、数据和信号处理。

［3］A well-designed service layer is capable to classify common application necessities and deliver APIs (application program interface) and protocols to sustenance required services, user needs, and applications. 设计良好的服务层能够对一般应用必需品进行分类，并提供API和协议以满足所需服务、用户需求和应用的需要。

［4］The services on the service layer run directly on limited network infrastructures to successfully find new services for an application, as they connect to the network. 为了成功地为一个应用程序找到新的服务，当连接到网络时，服务层的服务程序直接运行在有限的网络基础设施上。

［5］IoT could also importantly benefit by leveraging current Internet protocols such as IPv6, since this will make it possible to directly address any number of things required through the Internet. 通过利用现在的网络协议如IPv6，物联网受益颇多，因为这些协议将使它能够通过互联网直接处理所需的任何数量的设备。

[6] Other pros of the RFID system include offering detailed real-time information about the involved devices, dropping labor cost, shortening business process, enlarging the accuracy of inventory information, and improving business efficiency. RFID 系统的其他优点包括提供有关设备的详细实时信息，降低劳动成本，缩短业务流程，提高库存信息的准确性，提高经营效率。

[7] The first mentioned means services that represent the primary functionalities, which can be seen as the elementary service component and can be included by another service. 上面首先提到的是代表主要功能的服务，可以被看作是基本的服务组件，并可以包括在另一个服务中。

Unit 5

A The Conversion of Parts of Speech（词性的转换）

在英译汉过程中，有些句子可以逐词对译，有些句子则由于英汉两种语言的表达方式不同，就不能逐词对译，只能将词类进行转译之后，方可使译文显得通顺、自然。对词类转译技巧的运用需从以下四个方面加以注意：

1. 转译为汉语的动词

（1）名词转译为动词

英语中有些具有动作意义的名词，这些名词汉译时往往要转译为汉语的动词，译文才会通顺。

Example 1：Integrated circuits are fairly recent *development*.

译文：集成电路是近年<u>发展</u>起来的。

Example 2：A new nuclear power station is now in the process of *construction* in that area.

译文：那个地区正在<u>建设</u>一座新的核电站。

Example 3：In the dynamo, mechanical energy is used for *rotating* the armature in the field.

译文：在直流发电机中机械能是用来使电枢在磁场中<u>转动</u>的。

Example 4：Atomic energy is being used for the *production* of electrical power.

译文：原子能被用来<u>生产</u>电力。

（2）形容词转译为动词

当形容词做表语、主语、补足语或其他成分时，可以译为汉语的动词。

Example 1：Internet is *different* from intranet in many aspects though their spelling is *alike*.

译文：虽然拼写<u>相像</u>，internet 在很多方面<u>不同于</u> intranet.

Example 2：Are you *familiar* with the performance of this type of transistor amplifier?

译文：你<u>熟悉</u>这类晶体管放大器的性能吗？

Example 3：The design calculation will serve as an *illustrative* application of the theory of

PART I Techniques of EST E-C Translation（科技英语英译汉翻译技巧）

semiconductor devices.

译文：这些设计计算可用来证明半导体器件理论的实际应用。

（3）介词转译为动词

有些介词具有较强的动词意味，翻译时可以译为汉语动词。

Example 1：Noise figure is minimized *by* a parameter amplifier.

译文：采用参数放大器，即可将噪声系数减至最低。

Example 2：The company has advertised in the newspaper *for* electronics experts.

译文：公司在报上登广告招聘电子学方面的专家。

（4）副词转译为动词

英语中某些副词，如 on, off, up, in, out, over, behind, forward, away 等，在与系动词 be 构成合成谓语，或做宾语补足语，或状语时可以译为汉语的动词。

Example 1：In this case the temperature in furnace is *up*.

译文：在这种情况下炉温就会升高。

Example 2：When the switch is *off*, the circuit is open and electricity does not go through.

译文：开关断开时，电路开路，电流不能流过。

2. 转译为汉语的名词

无论是状态动词还是行为动词，在一定情况下均可转译为汉语的名词。

（1）动词转译为名词

Example 1：The design *aims* at automatic operation, simple maintenance and high productivity.

译文：设计的目的在于自动操作、维护简单和生产率高。

Example 2：This communication system is chiefly *characterized* by its ease with which it can be *maintained*.

译文：这种通信系统的主要特点是维修方便。

Example 3：The telecommunication *means* so much in modern life that without it our modern life would be impossible.

译文：电信在现代生活中的意义很大，没有它就不可能有我们现在的生活。

Example 4：The earth on which we live is *shaped* a ball.

译文：我们居住的地球，形状像一个大球。

（2）形容词转译为名词

当形容词做表语、主语、补足语或其他成分时，可以译为汉语的名词。

Example 1：IPC is more *reliable* than common computer.

译文：工控机的可靠性比普通计算机高。

Example 2：The steam turbine is less *economical* at cruising speed.

译文：汽轮机的巡航经济性较差。

Example 3：Television is *different* from radio in that it sends and receives a picture.

译文：电视和无线电的区别在于电视发送和接收的是图像。

（3）副词转译为名词

某些派生的副词可以转译为汉语的名词。

Example 1：This instrument is used to determine how *fully* the batteries are recharged.

译文：这种仪表用来测定电池的充电程度。

Example 2：The image must be *dimensionally* correct.

译文：图形的尺寸必须正确。

Example 3：Such magnetism, because it is *electrically* produced, is called electromagnetism.

译文：由于这种磁性产生于电，所以称为电磁。

Example 4：This device is shown *schematically* in Fig. 3-11.

译文：图 3-11 所示为这种装置的简图。

3. 转译为汉语的形容词

（1）副词转译为形容词

翻译时形容词和副词之间的转译最为常见。

Example 1：Some carbon compounds do react with oxygen rather *quickly*.

译文：某些碳化物同氧气起反应确实是相当快的。

Example 2：The sun affects *tremendously* both the mind and body of a man.

译文：太阳对人的身体和精神都有极大的影响。

Example 3：Java is *chiefly* characterized by its simplicity of operation and compatibility with almost all operation systems.

译文：Java 主要的特点是操作简单，并和几乎所有的操作系统兼容。

（2）名词转译为形容词

有些派生的名词翻译时可以直接翻译成与其相关的形容词。

Example 1：The electrical conductivity has great *importance* in selecting electrical materials.

译文：导电性在选择电气材料时是很重要的。

Example 2：Single crystals of high perfection are an absolute *necessity* for the fabrication of integrated circuits.

译文：高度完整的单晶体对于制造集成电路来说是绝对必要的。

Example 3：I am a *stranger* to the operation of this kind of machine.

译文：我对这类电机的操作是陌生的。

Example 4：The lower stretches of rivers show considerable *variety*.

译文：河流下游的情况是大不相同的。

4. 转译为汉语的副词

（1）形容词转译为副词

Example 1：The *same* principle of low internal resistance also apply to milliammeters.

译文：低内阻原理也同样地适用于毫安表。

Example 2：The modern world is experiencing *rapid* development of information technique.

译文：当今世界的信息技术正在迅速地发展。

Example 3：Transistors are *fairly recent* development.

译文：晶体管是最近才发展起来的。

Example 4：Electronic circuits are used for *fast and accurate* control of machines.

译文：我们用电子电路快速而准确地控制机器。

（2）名词转译为副词

有时根据上下文的含义，名词也可以翻译成副词，以使译文更加流畅。

Example 1：It is our great *pleasure* to note that our aeronautical industry is developing vigorously.

译文：我们<u>很高兴地</u>注意到我国的航空航天业正在蓬勃发展。

Example 2：The new mayor earned respect due to the *courtesy* of visiting the poor in the city.

译文：新市长<u>有礼貌地</u>访问城市贫民，赢得了人们的尊敬。

（3）动词转译为副词

Example：As a result, he *succeeded* in finding one substance that he could use for the filament.

译文：结果，他<u>成功地</u>找到了一种可以用来做灯丝的物质。

Exercises

Translate the following sentences into Chinese by the conversion techniques of the parts of the speech. Make sure to translate the underlined words or phrases properly.

（1）A body is negative charged when it has electrons in <u>excess</u> of its normal number.

（2）He was a good <u>calculator</u>, so we considered the answer correct.

（3）The <u>transformation</u> could be from any convenient input voltage to any convenient output voltage.

（4）If low-cost power becomes <u>available</u> from nuclear power plants, the electricity crisis would be solved.

（5）To some extend the smaller the assembly, the more <u>adaptable</u> is its manufacture to automated techniques.

（6）Power is needed to stall the armature <u>against</u> inertia.

（7）He had been <u>through</u> with the oiling of the parts before the machine was operated.

（8）An electron or an atom <u>behaves</u> in some ways as though it were a group of waves.

（9）Experiment indicates that the new chip is about 1.5 times as <u>integrative</u> as that of the old ones.

（10）Gold is an important metal but it is not <u>essentially</u> changed by man's treatment of it.

（11）A current varies <u>directly</u> as the voltage force and <u>inversely</u> as resistance.

（12）Concrete structures generally do not require much maintenance, <u>internally or externally</u>, if adequate provision is made for good expansion joints and firm foundations.

B GPS Vehicle Surveillance Equipment Is Here to Help You（全球卫星定位系统汽车监视装置正在帮助你）

Surveillance is not only a game of PI (private investigators) or the police. You can also have a surveillance gear (maybe not as sophisticated as wearable spy equipment or mini audio recorder measuring the size of your fingernail) to keep tracking almost anyone you want. What I'm talking about here is GPS tracking system.

GPS or global positioning system is a fully functional global navigation satellite system.[1] This system uses an artificial constellation of 24 medium Earth orbit satellites. These satellites transmit microwave signals, thus enabling a GPS receiver to determine its location, speed, direction and time. This system was developed by the United States Department of Defense and was named as NAVSTAR GPS which was given by Mr. John Walsh. NAVSTAR is not an acronym, as is widely believed.

GPS tracking systems are designed for tracking vehicle fleets, equipment, and people. GPS tracking system is used for fleet management, protecting the vehicle and driver, and locating equipment and people.[2] GPS tracking devices allow you to track your car or any other moving object, like boat, bike or even a plane. GPS trackers are even installed in some high-tech cell phones, which allow you to track a teenager or anyone who uses such a cell phone. However, the main use of a GPS tracking equipment is for vehicle surveillance.

How many times you've seen, heard, or maybe even experienced yourself how burglars steal a car right in the middle of the day. Such things happen every single day. And the worst part is that rarely do the police find the hijacked vehicle. So how can you protect yourself from such unpleasant "surprise"? That's right, by using a GPS tracking system.

But first, let's see where else GPS can be of good use.

1. GPS Can Be Installed Anywhere

Such GPS tracking devices are very small; they can even fit into the palm of your hand. So you can easily attach them to the bicycle or a motorbike for instance.

These devices can be easily attached to the car as long as they're magnetic. You could place one under the hood, in the back of your car or absolutely anywhere else you can think of. Moreover, these GPS trackers have waterproof casings, so you wouldn't have to worry about the damage of such devices.

GPS systems are used in yachts or boats by the owners or renting companies. As long as these gadgets are weather and water-proof, they do a great job for tracking the boat. They're very useful if you don't want to get lost.

Handheld GPS trackers are very good if you like trips. If you go to a picnic and suddenly get lost, you can be rest-assured that you'll find your way out, because these handheld GPS devices have easy to understand user interface and great mapping features.[3]

2. GPS Data for Vehicle Surveillance

As mentioned earlier, the main area where a GPS system is mostly used is for tracking vehicles. Such GPS devices are preferred by regular company owners or truck company owners.

Managers want to know who and when are abusing their cars. Because employees love to drive cars faster than they should. They love to drive the vehicle on weekends or holidays for other purposes than the work. So naturally, business owners want to track their employees and find out what's going on with their cars. And they have great options at doing this.

Monitor the Speed of the Vehicle—If an employee is driving too fast, he'll waste gasoline and ware tires off sooner that normally would. So a GPS tracker can help you in this situation easily. A GPS tracking system monitors the speed of the vehicle. Whenever the speed limit is exceeded, you get an alert and a report, so you'll have a proof and know who exceeded the speed.[4]

Know Where Your Car Is at the Moment—Let's say you're just turned on your computer and you want to login to the central station to find out where your car (or cars) is at the moment. GPS lets you do that without any trouble. You login to the system and see exactly where your car is now.

Signal Update Interval—You can set the desired update frequency of the signal of your GPS device. For instance, if you don't need to track your vehicles very often, you can set a frequency to 30 minutes. Otherwise, you can set it to 2 minutes and know where your car is precisely every 2 minutes.

GPS tracking system is very useful. Not only can it tell you where your car is located at the moment, it also tells you about the behavior of the person driving the car. Does he stop often near some burger restaurant? Does he take a break and spend some time in the bar? You'll know everything.

New Words and Expressions

surveillance　*n.*　监督，监视
GPS　全球定位系统（global position system）
medium-earth orbit satellite　中轨道卫星
fleet　*n.*　舰队
hood　*n.*　发动机罩
rest-assured　放心，确信无疑
find one's way out　寻找解决方法，寻找出路
mapping　*n.*　地图，绘图
alert　*n.*　警戒，警惕，警报
private investigator　私家侦探
navigation　*n.*　导航
constellation　*n.*　星座
gadget　*n.*　小玩意，小巧的机械装置
gasoline　*n.*　汽油
handheld　*adj.*　掌上型的，手持型的

Notes

［1］GPS or global positioning system is a fully functional global navigation satellite system. 全球定位系统（GPS）是一个全天候的全球导航卫星系统。

［2］GPS tracking system is used for fleet management, protecting the vehicle and driver, and locating equipment and people. GPS 追踪系统用于舰队管理、车辆和驾驶人的保护以及设备和人员的定位。

［3］If you go to a picnic and suddenly get lost, you can be rest-assured that you'll find your way out, because these handheld GPS devices have easy to understand user interface and great mapping features. 当你野餐突然迷路时，你能够确保找到解决途径，因为这些手持 GPS 设备有便于理解的用户界面和庞大的地图功能。

［4］A GPS tracking system monitors the speed of the vehicle. Whenever the speed limit is exceeded, you get an alert and a report, so you'll have a proof and know who exceeded the speed. GPS 追踪系统能监视车辆的速度。车辆超速时，你会接到提醒和报告，因此你就得到证据，知道是谁超速了。

C Adaptive Assistance: Smart Home Nursing（自适应帮助：智能家居护理）

1. Introduction

An important characteristic of smart technology is a seamless and implicit human computer interaction that uses wireless sensor/actuator devices to detect user situation and respond accordingly. In order to offer smart assistance, the system must have some means of assessing the context of interaction without explicit user intervention. This can be done by making both human's behavior and inner state a part of the processing loop, e. g. by deploying the sense-analyze-react principle performing a seamless observation, situation evaluation and active reaction. Having a generic support for such implicit and awareness-rich processing would allow deployment of smart technology in a whole range of medicare areas. [1]

Home health care consists of "a part-time skilled nursing care, physical therapy, occupational therapy, speech-language therapy, home health aide services, medical social services, durable medical equipment (such as wheelchairs, hospital beds, oxygen, and walkers) and medical supplies, and other services". Some of these services may be performed (in its basic form) without direct human participation. Recently, a new generation of control systems has been developed offering control strategy enriched by physiological and socio-behavioral analyses. Systems are called reflective as they diagnose users' physical, social and psychological state and react accordingly in a given situation. Such systems may significantly support home nursing, performing tasks where human presence is not necessary thus leaving more time for personal contact during the visits.

2. Theory Behind

This approach deploys the concept of a bio-cybernetic loop allowing for a multiple physiological sensing, composite analyses and decision making. The function of the loop is to monitor changes in user state in order to initiate an appropriate adaptive response. It takes results of affective computing and combines it with higher level understanding of social and goal-oriented situations. The approach is multi-modal as it takes into account different kinds of information, processing them in multiple loops at different time scales. There are three major phases of a single loop: sense, analyze and activate. These phases are repeated endlessly, where each consecutive cycle takes into account the effects of the previous one, performing constant self-tuning and self-optimization.

The first phase of the bio-cybernetic loop is monitoring of the user in a given situation. The collecting of information can be done by observing: ① overt actions (e. g. location, looking, pointing); ② overt expression (e. g. changes in behavior associated with psychological expression); and ③ covert expression (e. g. changes in physiology associated with psychological expression).

The analyses phase of the bio-cybernetic loop is a process that involves psychology, physiology and behavioral science knowhow. In order to make an effective use of such diverse data, this approach relays on affective and physiological computing results and deploys rule-based reasoning. The loop is designed according to a specific rationale, which serves a number of specific meta-goals, defined by the application needs.

The final phase of reaction is devoted to the adequate system response which is performed through a certain action of the system actuators having further influence on both the user and the controlled situation.[2]

Based on a better understanding of both personal involvement and social and behavioural situation of the user, a reflective system offers adaptive control of different types and different time scales: ① immediate adaptation supporting safety; ② short-term adaptation responding to a more complex states that require several steps of self-tuning; ③ long-term adaptation providing individualisation as a process that guarantees that the system has co-evolved with individual user and can target its functioning to specific individual needs.

3. Implementation

Developing reflective software involves tasks like real-time sensor/actuator control, user and scenario profile analyses, affective computing, self-organization and adaptation. To accomplish these requirements, a service and component-oriented middleware architecture, based on reflective ontology, has been designed that promises a dynamic and reactive behavior featuring different biocybernetic loops.[3] According to the reflective ontology, the reflective software is grouped into three layers: ① tangible layer—a low-level layer that controls sensor and actuator devices; ② reflective layer—with more complex services and components that evaluate user states; ③ application layer—a high level layer that defines application scenario and system goals; by combining low and high level services and components from other layers, application layer runs and controls the whole system.

The reflective layered architecture as a spin top, exercising different temporal loops at tangible, reflective and application level. A control loop (initialized with users' profile and scenario settings) starts by sampling the psycho-physiological and other measurements, continues with their analyses and finishes by adaptive system reaction. In a next iteration the system influence (caused by the reaction) can also be sensed and further tuned. The overall design goal is to have a generic modular structure that follows the patterns of immediate, short and long term adaptation and is capable of dynamic configuration and efficient functioning.

4. Application

Reflective technology has been successfully tested in the automotive domain implementing the concept of a "vehicle as a co-driver". Based on a comprehensive driver's psycho-physiological analyses and driving situation evaluation, the reflective vehicle features active support in driving by warning in case of high mental effort or dangerous driving situations, by adapting car entertainment according to driver's mood and by re-shaping the seat according to drivers comfort.[4] However, being genuinely generic and re-usable, reflective framework can be deployed in a range of different application domains.

A reflective home care for elderly people is currently under development. The goal is to construct a flexible "smart" ambient to control the home for elderly people offering both medical and rehabilitation services at one side and improving the quality of living at another.[5] The system should transform a home into friendly and supportive home nurse. The functionalities offered include:

1) Medical (physiological) monitoring;
2) Rehabilitation support;
3) Seating/laying comfort;
4) Monitoring of inhabitant movement;
5) Monitoring of home appliances;
6) Control of TV and media-rich entertainment;
7) Communication with mobile devices.

Enriched with reflective assistance, a smart home plays an active role in everyday life supporting medical diagnoses and check-ups, rehabilitation and physical exercises, watching health condition as well as emotional and mental state of the inhabitants.

Reflective support for each of the above mentioned functionalities is achieved by embedding numerous sensor and actuator devices into home settings and by deploying reflective control to these devices.[6]

8) Medical support consists of heart rate and blood pressure measurements—as regular daily check up, with warnings displayed on TV—and remote mobile monitoring by medical staff.

9) Rehabilitation support controls exercise devices (e.g. home cycle or walking track) according to the instruction displayed on television. Physiological monitoring (e.g. heart rate, blood pressure) is done simultaneously guiding the exercise according to the body response.

10) Comfort control is done via seating and laying sensors, checking the body pressure at critical points and modifying (if necessary) the shape of the armchair/matrices via air pumps.

11) Movements control is done via cameras placed at each room and is used only for critical hot line warnings (in case of sudden fall).

12) Home appliances are connected to automatic switch-on/off device allowing remote/mobile monitoring and control.

13) Entertainment control consists of a mood player designed to control entertainment center according to emotional response.

14) Communication control connects to mobile devices allowing for urgent calls and remote monitoring.

With the above described functionality, reflective home performs many of the routine functions of a home nurse. It does daily medical check-up, assists in active exercise and rehabilitation, cares for inhabitant physical, emotional and mental state, reminds and/or overtakes the control over switching on/off the home appliances and supports mobile communication with emergency center, friends and relatives. In summary it overtakes numerous medical, psychological, physical and social functions, complementing the work of a home nurse.

New Words and Expressions

seamless *adj.* 无缝的，无漏洞的
therapy *n.* 治疗

significantly *adv.* 意味深长地
bio-cybernetic *adj.* 生物控制的

PART I　Techniques of EST E-C Translation（科技英语英译汉翻译技巧）

self-tuning	n. 自调整，自适应	scenario profile analyses	场景概括分析
overt action	显性行动	tangible adj.	有形的
covert expression	隐性表达	configuration n.	配置
rationale n.	理论，基本说明	genuinely adv.	诚实地，真正地
meta-goal	元目标	rehabilitation n.	康复
individualization n.	个性化	routine adj.	日常的
co-evolved adj.	共同进化的		

Notes

［1］Having a generic support for such implicit and awareness-rich processing would allow deployment of smart technology in a whole range of medicare areas. 如果对这种隐式和感知丰富的处理有通用的支持，可以在整个医疗领域部署智能技术。

［2］The final phase of reaction is devoted to the adequate system response which is performed through a certain action of the system actuators having further influence on both the user and the controlled situation. 反应的最后阶段致力于充分的系统响应，这一响应是通过系统执行器的特定动作产生的，并进一步影响用户和控制状态。

［3］To accomplish these requirements, a service and component-oriented middleware architecture, based on reflective ontology, has been designed that promises a dynamic and re-active behavior featuring different biocybernetic loops. 为了实现这些要求，基于反射式本体，设计了这种服务和面向构件的中间件架构，以确保构成一个动态的具有不同生物控制回路特征的反应行为。

［4］Based on a comprehensive driver's psycho-physiological analyses and driving situation evaluation, the reflective vehicle features active support in driving by warning in case of high mental effort or dangerous driving situations, by adapting car entertainment according to driver's mood and by re-shaping the seat according to drivers comfort. 基于全面的驾驶人心理生理分析和驾驶状况评价，车辆反馈特征可以集中表现为高度脑力劳动或危险驾驶情况的警告驾驶，根据驾驶人的情绪改变娱乐设施，以及根据驾驶人的舒适感觉重塑座椅。

［5］The goal is to construct a flexible "smart" ambient to control the home for elderly people offering both medical and rehabilitation services at one side and improving the quality of living at another. 我们的目标是建立一个灵活的"智能"环境来控制老年人的家庭，一方面提供医疗和康复服务，另一方面也改善生活质量。

［6］Reflective support for each of the above mentioned functionalities is achieved by embedding numerous sensor and actuator devices into home settings and by deploying reflective control to these devices. 通过将许多传感器和执行器设备嵌入家庭环境中，并通过对这些设备进行反馈控制来实现对上述功能的回馈支持。

Unit 6

A　Amplification（增词）

英汉两种语言在遣词造句方面存在很大差异。有时为了使英文完整通顺需要采用增加词量的方法翻译，也就是在译文中增补一些在修辞上、语法结构上、语气上或语义上必不可少的词。在科技英语翻译过程中，通常有三种情况需要增加词量：① 有些词汇并没有在原文中出现，但是它们的意义已经暗含在句子中了；② 在原文中省略的词汇，如果译成汉语时也将其省略，译文会出现语句不连贯的情况；③ 英文词语意义表达清晰完整，但是逐词翻译出的汉语句子意义模糊，需要增加词量使译文完整。

1. 抽象名词的增译

英语中有许多抽象名词，这些词若直译，不能给人以具体明确的含义。因此，译成汉语时需要加上"情况""作用""现象""性""效果""方法""过程""设计""变化"等词语。

Example 1：*Oxidation* will make ground-wire of lightning arrester rusty.

译文：氧化作用会使避雷装置的接地线生锈。

Example 2：The *operation* of electric generators is based on this relationship between magnetism and electricity.

译文：发电机的工作原理是以磁和电这种相互作用为基础的。

Example 3：Electricity permits indication of *measurements* at a distance.

译文：电可以在相隔一段距离的地方显示各种测量结果。

2. 连词的增译

在英语中，不少句式中的因果、转折等连接关系并不都是通过逻辑性语法词来表示的，而是通过语法习惯或用词习惯来体现的。因此，翻译此类句子时应增译某些关联词。

Example 1：If the system gain is increased beyond a certain limit, the system will become unstable.

译文：如果系统的增益超过了一定界限，系统就会不稳定。

Example 2：The pointer of the ampere meter moves from zero to three and goes back to one.

译文：安培表的指针先从 0 转到 3，然后又回到 1。

Example 3：Were there no gravity, there would be no air around the earth.

译文：假如没有重力，地球周围就不会有空气。

3. 动词的增译

为了将英文原句意思表达完整，有时需要在英语名词前面或后面增加动词。

Example 1：He particularly stressed the need for an overall geological *survey* in the area.

译文：他特别强调要在这一地区进行一次地质普查。

Example 2: One way is to receive radio wave from space by means of relay satellite *in orbit*.
译文：一种方法是依靠在轨道上运行的转播卫星接收从太空来的无线电波。

Example 3: If we keep heating a *rod*, it will expand more as the temperature rises.
译文：如果我们把一根棒料不断地进行加热，随着温度的升高，棒料就会继续膨胀。

4. 省略词的增译

英语最忌讳重复，凡在前面出现的部分一般在后面便不再出现，以避免重复。但在翻译时往往需要将其增补上，这样才能使汉语语句完整。

Example 1: The letter *I* stands for the current in amperes, *E* the electromotive force in voltage, and *R* the resistance in Ohms.
译文：字母 I 代表电流的安培数，E 代表电动势的伏特数，R 代表电阻的欧姆数。

Example 2: Electrical energy can be turned into mechanical energy by an electric motor and into light energy in a lamp.
译文：电能可以由电动机转换成机械能，也可以由电灯转换成光能。

Example 3: This device keeps the air in the room at a higher pressure than outside and helps to stop air coming in.
译文：这个装置使室内的空气压力大于室外的空气压力，并阻止空气入内。

Example 4: The atmosphere moderates daytime temperature and retards night heat loss.
译文：大气层能缓和白天气温的上升和阻止夜间热量的损耗。

5. 概括性和分述性词语的增译

英语中在列举事实时很少用概括性的词语，在分述事物时也很少用分述性的词语，但汉语则恰恰相反。因此，翻译时需要根据情况增加概括性或分述性的词语。

Example 1: The article summed up the new achievements made in electronic computers, artificial satellites and rockets.
译文：本文总结了电子计算机、人造卫星和火箭这三方面的新成就。

Example 2: Based upon the relationship between magnetism and electricity are motors and generators.
译文：电动机和发电机是以磁和电这二者之间的关系为基础的。

Example 3: Both water waves and electromagnetic waves have the characteristics of velocity, frequency and amplitude.
译文：水波和电磁波都有速度、频率和波幅这三项特征。

Example 4: The frequency, wave length and speed of sound are closely related.
译文：声音的频率、波长和速度三者是密切相关的。

6. 解释性词语的增译

英语中某些词的词义有时隐含在其基本词义当中，翻译时若不将这种意思讲明，便不能准确地表达句意。

Example 1: A new kind of computer—cheap, small, light is attracting increasing attention.
译文：一种新型计算机正在引起人们越来越多的注意——它价格低，体积小，重量轻。

Example 2: Speed and reliability are chief advantages of IPC.
译文：速度快、可靠性高是工控机的主要特点。

Example 3: To every action there is always an equal and contrary reaction.

译文：对每一个作用力总是有一个大小相等、方向相反的反作用力。

7. 主语的增译

英汉翻译中，有时会将被动语态译成主动语态。这时英文中没有出现的行为主体便会在译文中出现，从而引起了增译。

Example 1: All bodies on the earth are known to possess weight.

译文：大家都知道地球上的一切物质都有重量。

Example 2: Rubber is found a good material for the insulation of cable.

译文：人们发现橡胶是一种用于电缆绝缘的理想材料。

Exercises

Translate the following sentences into Chinese by the amplification techniques.

(1) Integration can get rid of the static error in the closed-loop system.

(2) Without moving parts, maintenance requirements are cut to a minimum.

(3) Being stable in air at ordinary temperature, mercury combines with oxygen if heated.

(4) The printed circuits eliminated numerous hand operations and made possible circuits smaller than could be soldered by hand.

(5) Electricity is produced if acids react more with one metal than the other.

(6) Remember that a flow of electric charges in a coil forms an electromagnet. Also that a magnet induces a flow of charges.

(7) In spite of difficulties, the problem was overcome completely.

(8) During an earthquake, the vibration makes the earth's surface heave and tremble, and even crack open.

(9) Complicated as the problem is, it can be solved in an hour with the help of a computer.

(10) Pairs of shared electrons must spin in opposite directions.

B Introduction to Control Systems （控制系统介绍）

A control system is a device or set of devices to manage, command, direct or regulate the behavior of other devices or systems. There are two common classes of control systems, with many variations and combinations: logic or sequential controls, and feedback or linear controls. There is also fuzzy logic, which attempts to combine some of the design simplicity of logic with the utility of linear control. Some devices or systems are inherently not controllable. The term "control system" may be applied to the essentially manual controls that allow an operator, for example, to close and open a hydraulic press, perhaps including logic so that it cannot be moved unless safety guards are in place.

An automatic sequential control system may trigger a series of mechanical actuators in the correct sequence to perform a task. For example, various electric and pneumatic transducers may fold and glue a cardboard box, fill it with product and then seal it in an automatic packaging machine. In the case of linear feedback systems, a control loop, including sensors, control algorithms and actuators, is arranged in such a fashion as to try to regulate a variable at a setpoint or reference value. An example of this may increase the fuel supply to a furnace when a measured temperature drops. PID controllers are common and effective in cases such as this. Control systems that include some sensing of the results they are trying to achieve are making use of feedback and so can, to some extent, adapt to varying circumstances. Open-loop control systems do not make use of feedback, and run only in pre-arranged ways.

Control theory and control engineering deal with dynamic systems such as aircraft, spacecraft, ships, trains, and automobiles, chemical and industrial processes such as distillation columns and rolling mills, electrical systems such as motors, generators, and power systems, and machines such as numerically controlled lathes and robots. In each case the setting of the control problem is represented by the following elements:

1) There are certain dependent variables, calledoutputs, to be controlled, which must be made to behave in a prescribed way. [1] For instance, it may be necessary to assign the temperature and pressure at various points in a process, or the position and velocity of a vehicle, or the voltage and frequency in a power system, to given desired fixed values, despite uncontrolled and unknown variations at other points in the system. [2]

2) Certain independent variables, called inputs, such as voltage applied to the motor terminals, or valve position, are available to regulate and control the behavior of the system. Other dependent variables, such as position, velocity, or temperature, are accessible as dynamic measurements on the system.

3) There are unknown and unpredictabledisturbances impacting the system. These could be, for example, the fluctuations of load in a power system, disturbances such as wind gusts acting on a vehicle, external weather conditions acting on an air conditioning plant, or the fluctuating load torque on an elevator motor, as passengers enter and exit.

4) The equations describing the plant dynamics, and the parameters contained in these equations, are not known at all or at best known imprecisely. [3] This uncertainty can arise even when the physical laws and equations governing a process are known well, for instance, because these equations were obtained by linearizing a nonlinear system about an operating point. As the operating point changes so do the system parameters.

These considerations suggest the following general representation of the plant or system to be controlled. In Fig. 6-1 the inputs or outputs shown could actually be representing a vector of signals. In such cases the plant is said to be a *multivariable plant* as opposed to the case where the signals are scalar, in which case the plant is said to be a scalar or monovariable plant.

Control is exercised by feedback, which means that the corrective control input to the plant is generated by a device that is driven by the available measurements. [4] Thus, the controlled system can be represented by the feedback or closed-loop system shown in Fig. 6-2.

① Set to prescribed values called references;

② Maintained at the reference values despite the unknown disturbances;

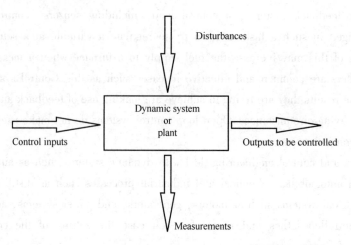

Fig. 6-1　A general plant

③ Conditions ① and ② are met despite the inherent uncertainties and change in the plant dynamic characteristics.

The first condition above is called *tracking*, the second, *disturbance rejection*, and the third, *robustness* of the system. The simultaneous satisfaction of ①, ②, and ③ is called robust tracking and disturbance rejection and control systems designed to achieve this are called robust servomechanisms.[5]

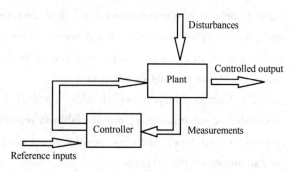

Fig. 6-2　A feedback control system

Notes

［1］There are certain dependent variables, calledoutputs, to be controlled, which must be made to behave in a prescribed way. 某些需要控制的因变量（叫作输出）必须使之以预定的方式运行。

［2］For instance, it may be necessary to assign the temperature and pressure at various points in a process, or the position and velocity of a vehicle, or the voltage and frequency in a power system, to given desired fixed values, despite uncontrolled and unknown variations at other points in the system. 尽管系统在其他点存在不可控的和未知的变量，但是可能仍然需要将工业过程中的温度和压力，或车辆的位置和速度，或电力系统的电压和频率配置到给定的理想固定值。

［3］The equations describing the plant dynamics, and the parameters contained in these equations, are not known at all or at best known imprecisely. 描述系统动态的方程和方程中包含的参数是完全未知的，至少是不精确的。

［4］Control is exercised by feedback, which means that the corrective control input to the plant is generated by a device that is driven by the available measurements. 控制由反馈实现，这意

味着被控对象的校正控制输入由可测变量驱动的装置产生。

[5] The first condition above is called *tracking*, the second, *disturbance rejection*, and the third, robustness of the system. The simultaneous satisfaction of ①, ②, and ③ is called robust tracking and disturbance rejection and control systems designed to achieve this are called robust servomechanisms.

第一个条件称作系统的跟踪能力，第二个条件称作系统的抗干扰性，第三个条件称作系统的鲁棒性。同时满足①、②和③称作鲁棒跟踪和干扰抑制，满足以上条件的控制系统称作鲁棒伺服机构。

New Words and Expressions

trigger　　*v.*　扳动扳机射击，触发
　　　　　n.　（枪上的）扳机，触发器，引爆器
cardboard　*n.*　硬纸板
monovariable　*n.*　单变量

distillation　*n.*　蒸馏
sequential　*adj.*　顺序的
velocity　*n.*　速度
hydraulic　*adj.*　液压的

C　IoT Challenges and Future Directions（物联网的挑战与未来发展）

During the last 10 years, Google research trends have been measuring the number of times that the IoT key has been checked in Google navigator. They clearly noticed that the IoT research is constantly increasing. The IoT is also expected to take 5-10 years to be adopted by the market. Technically, the trend of the IoT is the incorporation of sensing and the Internet. All the networked things should be flexible, smart, and autonomous enough to provide required services. It will provide our daily lives services with desired connectivity and intelligence.

1. Architecture and Dependencies

In the presence of a huge number of objects which are connected to the Internet, it is indispensable to have an adequate architecture that supports and permits an easy and large connectivity, tight control, ideal communication, and useful applications.[1] Designing an SOA for the IoT is a big challenge for a large scale network. The automatic service composition based on the requirements of applications is still a challenge as well. To realize the services' compatibility in different implementation environments, a commonly accepted service description language is required. Also, a robust service discovery is essential to progress the IoT technology. Since the app store can offer novel and unbounded development applications that can simply be executed on smart phones, they may be exploited and used as an architectural approach for the IoT. Guabbi et al. present an architecture that suits some IoT application domains. It is based on cloud computing, where cost based services are required. The European Union projects of SENSE and the Internet of things-architecture (IoT-A) have been basically focusing on challenges from the WSN aspect for being the prevalent component in the IoT. Generally, an overall architecture needs to be tightly developed according to the needs and functionalities of fields so as it can supply the necessary network and information management services to empower a reliable and accurate context information retrieval and actuation on the physical environment[2].

It is noted that the IoT systems are not complicated, but designing and building them can be a complicated task. Different applications with shared sensor and actuator devices may have to run simultaneously to build a complete, reliable, and synchronized system. If the system is not tightly and carefully designed, systems of systems interference problems may occur. In a hyper-connected world, if an error occurs in just one part of a system, this can cause a dislocation in the whole system, which may turn the human life into chaos. To alleviate such a serious problem, a comprehensive approach for specifying, detecting, and resolving dependencies across applications should be considered while designing the IoT architecture. Also, in hyper-connected IoT systems, a reduction of the complexity of connected systems, the enhancement of the security, standardization of applications, and the guarantee of the safety and privacy of users must be available anytime, anywhere, and on any device.

2. Security

The frequent major problem that can really negatively impact the network systems is security attacks. Security attacks can easily affect the IoT systems for many reasons, such as the minimal capacity of the smart devices, the wireless communication between devices, the openness of systems, and the physical accessibility to sensors, actuators, and objects. Hence, RFID, WSN and clouds are the three IoT hardware that might be subjected to such attacks. It is noticed that RFID is the most vulnerable device as it identifies and tracks persons as well as objects. Devices are notoriously prone to failure. Once a device is broken, it can be easily exploited by attackers. In this case, the system with high fault tolerance must be deployed. The redundancy of devices enables the system to continue operating properly even when there is a device failure. It is mentioned that 70% of the IoT devices are vulnerable and are prone to any kind of malware. This problem is prevailing on account of the lack of transport encryption, insecure Web interfaces, inadequate software protection, and insufficient authorization. To have a reliable system, its data must not be altered. The nodes in the system must be authorized to send or receive data streams. The message authentication codes guarantee the integrity and authenticity. So, the authentication is a critical and important factor that the system must ensure. The authentication problem can be found in the traditional network reprogramming, which just consists of a data dissemination protocol that distributes code to all the nodes in the network without authentication, and this is a threat that affects security.[3] To achieve a secure programming protocol, the device must be aware about the authenticity of every code to prevent any malicious installation.

To address the security attack problems and to let the IoT system work properly, the IoT system needs to detect the attack, diagnose it, and deploy countermeasures and repairs. In other words, it must also adapt to new, unexpected attacks, especially when the system is first deployed. As most of the IoT's devices are resource-constrained devices, these procedures must be performed in a lightweight manner. To have a reliable system, all these procedures must be performed using real-time computation. Procedures that train developers to integrate security solutions (e.g., intrusion prevention systems, firewalls) are needed.

To gain the social acceptance of the IoT technologies, the trustworthiness of information and protection of private data must be tightly verified. The data provenance and integrity, identity management, trust management, and privacy are the main challenges in building a secure IoT architecture. These points are so critical as data provenance ensures that the source of data is trustworthy, which refers to the trustworthiness of the system. Data integrity ensures the absence of unauthorized alteration of the data. Trust management ensures trust in the devices or embedded systems. Identity management refers to the administration of individual identities. Kanuparthi et al. outlined the main different attacks that can affect these four main challenges. They proposed appropriate solutions for each attack.

Challenge 1: Data provenance and integrity. The first attack that can affect the data provenance and data integrity is when the physical sensor is maliciously modified to send incorrect values. The sensor physical unclonable functions (PUF) is considered as a perfect solution as it provides authentication, unclonability, and verification of a sensed value.

Challenge 2: Identity management. A malicious node can falsify its identity into a genuine one to be authorized to send data, to the node/sensor or the actuator to carry out some unnecessary actions. The proposed solution for such problem is sensor PUFs. The sensor PUFs have the ability to supply unique IDs. A lightweight identity management application that performs the necessary tasks is required.

Challenge 3: Trust management. The applications of the IoT's devices must be protected against attackers. Also, the actuators must be protected against the inputs generated from unauthorized sensors. The attackers can affect the application by secretly executing a rootkit. The proposed solution for this case is the implementation of the hardware performance counters (HPCs). The HPCs are registers that can monitor certain events that occur during the lifetime of a program so that they can detect the integration of the rootkit program.

3. Privacy

The mobile crowd sensing and cyber-physical cloud computing IoT technologies contribute in the emergence of the smart and connected communities (SCC). The goal of building SCC is to live in the present, plan for the future, and remember the past. Since SCC are human-centered systems, security and privacy are considered as a major concern that must be seriously addressed. For example, in order to solve traffic problems, GPS sensor readings can be used to estimate traffic congestion. On the other hand, they can be also used to show private information about people (e. g. home, work locations, etc.).

The privacy problem can occur when two devices communicate with each other. The attacker can recover the private information transformed from the sender to the receiver or vice versa. As the cryptographic methods are intended to protect the system against outside attackers and ensure data confidentiality, lightweight encryption algorithms are the ideal solution to guard against the eavesdropping attack and to ensure the confidentiality of the data.[4] Also, the existing encryption technology used in WSNs can be extended and deployed in the IoT, such as encryption algorithms, hash functions, digital signatures, and key exchange algorithms. The selection of the appropriate algorithm should be according to the IoT's capacity constraint like power and memory. The possibility of the interconnection of a myriad number of devices in the IoT system is also a reason that violates the privacy. To respect the devices' privacy, the definition privacy from the social, legal, and cultural perspectives is needed. Also, a set of dynamic privacy policies must be deployed. In the case of an incoming user request, an evaluation of the request must be applied. If the users' request respect the established policies, the request will be granted; otherwise, it will be denied. In this case, the IoT paradigm must be able to express users' requests for data access and the policies. The main issue is the need for the creation of a new language that expresses privacy policies. This lack is caused by the difficulty of the interpretation of many requirements.

4. Openness

Openness matters a lot in such a broad scale network as it makes it easy to remotely control devices by sending data over the Internet. Another great benefit is the cooperation and 2-way control. It is noticed that the benefits cannot be achieved except in the presence of the open communication systems, where several objects can constantly communicate with each other. On the other hand, openness creates many new research problems that must be considered while designing the system, such as security and privacy. Feedback control is also one of the problems that openness encounters. Usually sensors and actuators use feedback control theory to provide robust performance. Studies have proven that stochastic control, robust control, distributed control, and adaptive control are not developed well enough, so a rich set of techniques must be employed.

5. Standardization

Standardization is considered as a main part that aids the development of IoT. However, rapid emergence of IoT makes the standardization difficult. The common issues of the IoT standardization are: radio access level issues, semantic interoperability and security, and privacy issues. Moreover, APIs are also quite demanded in the interconnection among heterogeneous smart objects. The web based interfaces are considered as the perfect solution that facilitates these communications, especially with the IoT devices and cloud computing while transmitting data for storage.[5] The main issue resides in the absence of the specific design for an efficient machine-to-machine communication. The open standards of the IoT, such as security standards, communication standards, and identification standards, might empower the growth of the IoT technologies.

6. New Protocols

The protocols play a main role to ensure the exchange data between devices and the efficient transport of embedded devices' data to the Internet. In familiar web protocols and in several MAC protocols there have been so many proposed protocols for various domains with time division multiple access (TDMA), carrier sense multiple access (CSMA) and frequency division multiple access (FDMA). Although, all these protocols cannot suit the IoT devices well. So, to achieve acceptable results, new lightweight protocols must be redefined taking into account the constraints of the device's energy. Sensors are vulnerable. Therefore, they can be out of order at any moment and for any reason. To build a reliable system, the latter must be capable to perform as a self-adapting network and to allow for multi-path routing. The number of hops in the multi-hop scenario will be limited. For that, routing protocols must be modified to take into consideration the energy constraints of the device.

7. Energy

The huge number of heterogeneous simultaneous sensing of the urban environment negatively affects the network traffic, data storage, and energy utilization. The reduction of energy consumption is a critical point due to the limited power of the IoT devices. So, a set of techniques that are capable to improve energy efficiency are required. Compressive sensing enables one to reduce the signal measurements without having an effect on the accurate reconstruction of the signal. Using sampling methods, the system can lower the data transmission rate. Also, the transmission power of each sensor can be reduced due to the synchronous communication used by compressive wireless sensing (CWS).

8. Extracting Knowledge from Big Data

In the IoT, the quantity of the generated data from the myriad number of smart objects disseminated on planet earth is growing exponentially. It is about streaming data that gives information about location, movement, vibration, temperature, humidity, and chemical changes in the air, etc. In the era of IoT, the SCC deploys wireless sensors that may cover diverse application domains such as, healthcare, environmental, smart buildings, and smart interconnected automobiles and trucks. The main problem is how to translate physical, biological, or social variables into a meaningful electrical signal. To handle the data heterogeneity, SUN et al. proposed to address the following issues: How to improve quality (accuracy) of data in real time; how to improve intelligent data interpretation and semantic interoperability; how to unify data representation and processing models to accommodate heterogeneous or new types of data; how to implement inter-situation analysis and prediction; how to implement knowledge creation and reasoning; and how to conduct short-term and long-term storage.[6]

Big data stream mobile computing (BDSMC) is a new paradigm emerged from the convergence of broadband mobile Internet networking (B-MINet) and real-time mobile cloud computing (Rt-MCC). The BDSMC aims at describing a new generation of mobile/wireless integrated computing-networking infrastructures. Baccarelli et al. proposed "five Vs" formal characterization of the BDSMC paradigm, which are: volume (ever increasing amount of data to be processed), velocity (data generation at fast and unpredictable rates), value (huge value but hidden in massive datasets at very low density), and volatility (the acquired data streams must be transported and processed in real time).[7] They described the volatility as 5th V. The volatility is used in featuring big data stream applications. According to these 5 Vs characterization, the life cycle of big data streams can be composed of three phases: data acquisition and local preprocessing at the mobile devices, data transport, and data post-processing at the remote data centers. To gain user trust, a tight interpretation is required. To create a valuable knowledge from a large qualitative data, data mining techniques must be intervened. The main challenge is the extraction of useful information from a complex sensing environment. For that reason, advanced data mining tools are needed. Data needs to be understood using computational and mathematical models. The shallow learning methods are those that are most currently used. They attempt to make sense of data by using supervised and unsupervised learning for extracting pre-defined events and data anomalies. The major research problem in complex sensing environments is how to simultaneously learn representations of events and activities at multiple levels of complexity. Deep learning, the branch of machine learning is based on algorithms that attempt to model multiple layers of abstraction that can be used to interpret given data. Deep learning has to construct adaptive algorithms, distributed, and incremental learning techniques responding to resource constraints in sensor networks. To realize a reliable interpretation and modeling knowledge to get good comprehensible and usable knowledge, many challenges must be treated. When addressing noise, a reliable signal processing technique must be applied to obtain a pure signal. Developing new and efficient inference techniques must be realized. As some of the same data streams can be used for many different purposes, data issuers and how the data was processed must be known.

New Words and Expressions

robust　*adj.*　强健的，健康的，粗鲁的
IoT-A（the Internet of things-architecture）物联网的体系结构
retrieval　*n.*　检索
synchronized　*adj.*　同步的
vulnerable　*adj.*　脆弱的
notoriously　*adv.*　著名地，众所周知地
integrity　*n.*　完整，诚实
authenticity　*n.*　可靠性，真实性
provenance　*n.*　起源，出处
PUF（physical unclonable functions）物理不可克隆功能
malicious　*adj.*　恶意的
HPCs（hardware performance counters）硬件性能计数器
SCC（smart and connected communities）
encryption　*n.*　加密
hash function　散列函数
proven　*adj.*　经过验证或证实的

TDMA（time division multiple access）时分多址
CSMA（carrier sense multiple access）载波侦听多址接入
FDMA（frequency division multiple access）频分多址
heterogeneous　*adj.*　各种各样的，成分混杂的
CWS（compressive wireless sensing）压缩无线传感
BDSMC（big data stream mobile computing）大数据流移动计算
B-MINet（broadband mobile Internet networking）宽带移动互联网
Rt-MCC（real-time mobile cloud computing）实时移动云计算
shallow　*adj.*　肤浅的，表面的
incremental　*adj.*　增加的

Notes

[1] In the presence of a huge number of objects which are connected to the Internet, it is indispensable to have an adequate architecture that supports and permits an easy and large connectivity, tight control, ideal communication, and useful applications. 在大量对象与因特网相连的情况下，必须有一个足够的体系结构来支持和允许简单而庞大的连接、严密的控制、理想的通信和有用的应用程序。

[2] Generally, an overall architecture needs to be tightly developed according to the needs and functionalities of fields so as it can supply the necessary network and information management services to empower a reliable and accurate context information retrieval and actuation on the physical environment. 通常，根据现场需求和功能，需要研发出一个整体架构，以使其能够支持必需的网络和信息管理服务，去促成在物理环境中的可靠而准确的语境信息检索和驱动。

[3] The authentication problem can be found in the traditional network reprogramming, which just consists of a data dissemination protocol that distributes code to all the nodes in the network without authentication, and this is a threat that affects security. 传统的网络重新编程中存在认证问题，它只包含一个数据分发协议，它不经身份验证就向网络中的所有节点分发代码，这是影响安全性的一种威胁。

[4] As the cryptographic methods are intended to protect the system against outside attackers and ensure data confidentiality, lightweight encryption algorithms are the ideal solution to guard against the eavesdropping attack and to ensure the confidentiality of the data. 由于加密方法是为了保护系统免受外部攻击者的攻击并确保数据的保密性，因此简单的加密算法是防止窃听攻击和确保数据机密性的理想解决方案。

[5] The web based interfaces are considered as the perfect solution that facilitates these communications, especially with the IoT devices and cloud computing while transmitting data for storage. 基于网络的界面可以看作是实现这些通信的完美方案，特别是在为保存而传输数据时使用物联网设备和云计算的情况下。

[6] To handle the data heterogeneity, SUN et al. proposed to address the following issues: How to improve quality (accuracy) of data in real time; how to improve intelligent data interpretation and semantic interoperability; how to unify data representation and processing models to accommodate heterogeneous or new types of data; how to implement inter-situation analysis and prediction; how to implement knowledge creation and reasoning; and how to conduct short-term and long-term storage. 为了解决数据异构性，Sun 等人提出要解决以下问题：如何实时提高数据的质量（精度）；如何提高智能数据解释和语义互操作性；如何统一数据表示和处理模型以适应异构或新的数据类型；如何实现内部情境分析和预测；如何实现知识的创造和推理；如何进行短期和长期存储。

[7] Baccarelli et al. proposed "five Vs" formal characterization of the BDSMC paradigm, which are: volume (ever increasing amount of data to be processed), velocity (data generation at fast and unpredictable rates), value (huge value but hidden in massive datasets at very low density), and volatility (the acquired data streams must be transported and processed in real time). 巴卡莱利等人提出了用"5V"表征大数据流移动计算范式，即体积（不断增加的、要处理的数据量）、速度（快速且不可预测的速率进行数据生成）、价值（隐藏在海量稀疏数据中的巨大价值）和波动性（采集到的数据流必须实时传输和处理）。

Unit 7

A Omission（减词）

汉译英中，与增词译法相似，有时为了使译文符合汉语的规范和习惯，应将一些汉语中不应出现的词省略不译，这就是省译法。省译法的原则是省词不省意，省译后应在内容上与原文保持一致。

1. 冠词的省译

英语中的冠词，是古英语中的指示代词和前置于名词的形容词的衍生词，属于虚词。它的基本功能是语法的，不是词义的，其本身并没有独立的词汇意义。在现代英语中，冠词只是起一种指示范围的作用，翻译中在多数情况下是应该省译的。

Example 1：The controlled output is *the* process quantity being controlled.

译文：被控输出量是指被控的过程变量。

Example 2：A transistor is *a* device controlling the flow of electricity in a circuit.

译文：晶体管是控制电路中电流的器件。

Example 3：Sensor switches are located near *the* end of the feed belt.

译文：传感器开关位于给料带的末端。

2. 代词的省译

英语中的代词有人称代词、物主代词、反身代词和关系代词等，翻译时没有特指时也是可以省略的。

Example 1：Different metals differ in *their* conductivity.

译文：不同的金属具有不同的导电性能。

Example 2：The current will blow the fuses when *it* reaches certain limit.

译文：当电流达到一定界限时会使熔丝熔断。

Example 3：*It* seems that the result of this test will be very important to us.

译文：看来，这个试验结果对我们非常重要。

Example 4：The development of IC made *it* possible for electronic devices to become smaller and smaller.

译文：集成电路的发展使电子器件可以做得越来越小。

Example 5：The energy *that* has to be supplied by the generator in order to overcome the opposition is transformed into heat within the conductor.

译文：发电机克服电阻所必须发出的能量在导体内转换成了热。

Example 6：The gas distributes *itself* uniformly throughout a container.

译文：气体均匀地分布在整个容器中。

Example 7：Insulators can also conduct electricity to some extent, though *they* give a high resistance to an electric current.

译文：虽然绝缘体对电流呈很大电阻，但多少也有一点导电性。

3. 动词的省译

英语的各类句型有一个共同的特点，就是必须有谓语动词（省略句另当别论）。而汉语则不然，除了动词以外，形容词等也可以做谓语，因此翻译成汉语时动词有时可以省略。

Example 1：The electric power industry *is* primarily concerned with energy conversion and distribution.

译文：电力工业与能源的转化和分配密切相关。

Example 2：The importance of Internet *becomes* known to all day by day.

译文：因特网的重要性日渐被所有人认识。

Example 3：This diode *produces* about nine times more radiant power than that one.

译文：这只二极管的耗散功率比那只大 8 倍。

Example 4：Scientists *have experimented* with attempts to turn fog into light rain by ultrasonic means.

译文：科学家试图用超声波把雾变成小雨。

Example 5：Pentium CPU *provides* a relatively high speed than X86.

译文：奔腾 CPU 的运行速度比 X86 相对要快。

Example 6：A CD *has* a much more data volume than that of a hard disk or a floppy disk.

译文：光盘的数据存储量比硬盘或软盘大得多。

Example 7：The output of transformer *has* a voltage of 10 kilovots.

译文：变压器的输出电压为 10kV。

Example 8：When heated, gases act in exactly the same way as liquid *acts*.

译文：气体受热时发生的变化与液体完全一样。

4. 介词的省译

介词在英语中频繁使用，但在翻译时也是经常省略不译的。

Example 1：The notebook computer carries *with* it the storage battery of itself.

译文：笔记本式计算机自带蓄电池。

Example 2：A compact disk is a good storage medium with a volume *of* several hundred megabytes.

译文：光盘是一种很好的存储介质，其存储容量达几百兆字节。

Example 3：The search *for* even better magnetic materials is a part of the modern front of physics.

译文：寻找更好的磁性材料是现代物理学的前沿问题之一。

Example 4：They found the automatic production line *in* proper operation.

译文：他们发现自动生产线运转正常。

Example 5：A strain means either a change *in* size or a change *in* shape.

译文：张力就意味着改变大小或形状。

Example 6：Space is not empty, but contains great number of fragments *from* a shooting star.

译文：空间并不是空荡荡的，而是含有大量的流星碎片。

5. 连词的省译

英语中的连词表示了词之间或上下文之间的逻辑和语法关系。而汉语中的这些关系主要是由词序来体现的，其彼此之间的逻辑关系往往是内含的，因此翻译时连词常常省略。

Example 1：Practically all substances expand *when* heated and contract *when* cooled.

译文：实际上一切物质都是热胀冷缩的。

Example 2：There are two cylinders of equal mass, one being solid *and* the other hollow.

译文：有两个质量相等的圆柱体，一个是实心的，另一个是空心的。

Example 3：The experiment leads us to the discovering *that* electricity, heat, light and magnetism are all forms of energy.

译文：这些实验使我们发现，电、热、光和磁都是能量的形式。

Example 4：The average speed of all molecules remains the same *as long as* the temperature is constant.

译文：温度不变，分子的平均运动速度就不变。

6. 同义词的省译

英语中有一物多名的现象，当英语句子出现多个同义词或近义词表示同一事物或同一语义，而汉语中与其对应的词汇只有一个时，翻译时需要省译，即对多个同义词的相同含义只翻译一遍。

Example 1：Electric *generators*, or *dynamos*, are most useful of all electrical devices.

译文：发电机是一切电器装置中最有用的装置。

Example 2：To be sure, the change of the earth is slow, *but*, *nevertheless*, it is continuous.

译文：确实，地球的变化是缓慢的，但这一变化却是连续的。

Example 3：It is essential that the *mechanic or technician* understand well the characteristics of battery circuits and the proper methods for connecting *batteries and cells*.

译文：重要的是技术人员要深入了解电池电路的特性和连接电池的恰当方法。

Example 4：Semiconductor devices have no *filaments or heaters* and therefore require no heating power or warming up time.

译文：半导体器件没有灯丝，因此不需要加热功率或加热时间。

Exercises

Translate the following sentences into Chinese by the omission techniques. Make sure to translate the underlined words or phrases properly.

（1）The behavior of <u>the</u> motion of <u>the</u> molecules in <u>a</u> gas and in a liquid is different from each other.

（2）Friction always manifests <u>itself</u> as a force <u>that</u> opposes motion.

（3）The cerebrum is the place where man uses what <u>he</u> has learned about <u>his</u> environment so that <u>he</u> can master and control it.

（4）We consider <u>it</u> quite necessary to monitor conditions existing on the system continuously.

（5）Residential loads on electric power systems are ordinarily supplied by circuits <u>that</u> normally operated at 120 – 240 volts.

（6）This laser beam <u>covers</u> a very narrow range of frequencies.

（7）Like charges repel each other <u>while</u> opposite charges attract.

(8) Rapid oxidation accompanied by the liberation of noticeable heat and light is the process of <u>combustion</u>, or <u>burning</u>.

(9) Steel and iron products are often rusted if <u>they</u> are located in moist places.

(10) If the chain reaction is out of control, <u>it</u> would cause a terrible explosion.

(11) Series circuits do have the advantage of increasing resistance to reduce current when <u>this</u> is desirable.

(12) The radio telescope has enabled astronomers to study <u>the</u> matter between <u>the</u> stars in what was once thought of as simply space.

B Design of Control Systems（控制系统设计）

Starting with the controlled process such as that shown by the block diagram in Fig. 7-1, control system design involves the following three steps:

1) Determine what the system should do and how to do it (design specifications).

2) Determine the controller or compensator configuration, relative to how it is connected to the controlled process.

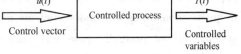

Fig. 7-1 Controlled process

3) Determine the parameter values of the controller to achieve the design goals.

We often use design specifications to describe what the system should do and how it is done. These specifications are unique to each individual application, and often include specifications about relative stability, steady-state accuracy (error), transient-response, and frequency-response characteristics. In some applications there may be additional specifications on sensitivity to parameter variations, that is, robustness, or disturbances rejection.

The design of linear control systems can be carried out in either the time domain or the frequency domain. For instance, steady-state accuracy is often specified with respect to a step input, a ramp input, or a parabolic input, and the design to meet a certain requirement is more conveniently carried out in the time domain.[1] Other specifications such as maximum overshoot, rise time, and settling time, are all defined for a unit-step input, and therefore are used specifically for time-domain design. We have learned that relative stability is also measured in terms of gain margin, and phase margin. These are typical frequency-domain specifications, which should be used in conjunction with such tools as Bode plot, polar plot, gain-phase plot, and Nichols chart.[2]

We have shown that for a second-order prototype system, there are simple analytical relationships between some of these time domain and frequency-domain specifications.[3] However for higher-order systems, correlation between time-domain and frequency-domain specifications are difficult to establish. As pointed out earlier, the analysis and design of control systems is pretty much an exercise of selecting from several alternative methods for solving the same problem. Thus, the choice of whether the design should be conducted in the time domain or the frequency domain depends often on the preference of the designer. We should be quick to point out, however, that in most cases, time-domain specifications such as maximum overshoot, rise time, and settling time are usually used as the final measure of system performance. To an inexperienced designer, it is difficult to comprehend the physical connection between frequency-domain specifications such as gain and phase margins, and resonance peak, to actual system performance. For instance, does a gain margin of 20 db guarantee a maximum overshoot of less than 10 percent? To a designer it makes more sense to specify, for example, that the maximum overshoot should be less than 5 percent, and a settling time less than 0.01sec. It is less obvious what, for example, a phase margin of 60° and an M_r of less than 1.1 may bring in system performance. The following outline will hopefully clarify and explain the choices and reasons for using time-domain versus frequency-domain specifications.

1) Historically, the design of linear control systems was developed with a wealth of graphical tools such as Bode plot, Nyquist plot, gain-phase plot, and Nichols chart, which are all carried out in the frequency domain. The advantage of these tools is that they can all be sketched by following approximation methods without detailed plotting. Therefore, the designer can carry out designs using frequency-domain specifications such as gain margin, phase margin, M_r and the like. High-order systems do not generally pose any particular problem. For certain types of controllers, design procedures in the frequency domain are available to reduce the trial-and-error effort to a minimum.

2) Design in the time domain using such performance specifications as rise time, delay time, settling time, maximum overshoot, and the like, is possible analytically only for second-order systems, or for systems that can be approximated by second-order systems. General design procedures using time-domain specifications are difficult to establish for system with order higher than the second.

3) The development and availability of high-powered and use-friendly computer software is rapidly changing the practice of control system design, which until recently had been dictated by historical development. [4] With modern computer software tools, the designer can go through a large number of design runs using the time-domain specifications within a matter of minutes. This diminishes considerably the historical edge of the performing graphical design manually.

Finally, it is generally difficult (except for an experienced designer) to select a meaningful set of frequency-domain specifications that will correspond to the desired time-domain performance requirements. [5] For example, specifying a phase margin of 60° would be meaningless unless we know that it corresponds to a certain maximum overshoot. As it turns out, to control maximum overshoot, usually one has to specify at least phase margin and M_r eventually, establishing an intelligent set of frequency-domain specifications becomes a trial-and-error effort.

However, frequency-domain methods are still valuable in interpreting noise rejection and sensitivity properties of the system, and, most important, they offer another perspective to the design process. Therefore, the design techniques in the time domain and the frequency domain are treated side by side, so that the methods can be easily compared.

New Words and Expressions

steady-state accuracy (error)　稳态精度（误差）　　step input　阶跃输入

ramp input	斜坡输入	time-domain *n.*	时域
parabolic input	抛物线输入	frequency-domain *n.*	频域

Notes

［1］For instance, steady-state accuracy is often specified with respect to a step in put, a ramp input, or a parabolic input, and the design to meet a certain requirement is more conveniently carried out in the time domain. 例如，稳态误差通常是对应于阶跃输入、斜坡输入或者加速度输入要求的，为满足某些需要的设计，在时域中实施比较方便。

［2］We have learned that relative stability is also measured in terms of gain margin, and phase margin. These are typical frequency-domain specifications, which should be used in conjunction with such tools as Bode plot, polar plot, gain-phase plot, and Nichols chart. 我们知道相对稳定性也可以用增益裕度和相位裕度衡量，它们是典型的频域指标，应该和伯德图、极坐标图、幅相曲线和尼古拉斯图等工具一起配合使用。

［3］We have shown that for a second-order prototype system, there are simple analytical relationships between some of these time domain and frequency-domain specifications. 已经指出，对于一个二阶模型系统，时域指标和频域指标间有简单的解析关系。

［4］The development and availability of high-powered and user-friendly computer software is rapidly changing the practice of control system design, which until recently had been dictated by historical development. 控制系统的设计实现一直受到历史方法的影响，直到最近高性能和便于操作的计算机软件的发展和普及正飞速地改变着它的方法。

［5］Finally, it is generally difficult (except for an experienced designer) to select a meaningful set of frequency-domain specifications that will correspond to the desired time-domain performance requirements. 最后，选择一组有意义的和希望的时域运行指标相匹配的频域指标是有困难的（除非是经验丰富的设计人员）。

C The Internet of Things in Manufacturing: Benefits, Use Cases and Trends（制造业的物联网：优点、实例和发展趋势）

In the age of Industry 4.0 and the digital transformation of manufacturing, the manufacturing industry is the market where most industrial Internet of things (IIoT) projects are realized and by far the market where most IIoT investments are made.

IoT is a core component of industrial transformation efforts across the globe, including Industry 4.0 (with its fourth Industrial Revolution) and the industrial Internet (with the Industrial Internet Consortium). Moreover, manufacturing isn't just the clear leader in the industrial Internet but it tops ALL industries (including the consumer IoT space) in the broader Internet of things reality. According to IDC data, published early 2017, the manufacturing industry was good for a total IoT spend of $178 billion in 2016, which is more than twice as much than the second largest vertical market (in IoT spend), transportation.

1. IoT Spend and Evolutions in Manufacturing 2017 – 2020

Manufacturing is poised to keep that first position across the globe until at least 2020, even if, as usual, in some geographies, this leading position is more outspoken.

In the APeJ (Asia Pacific, excluding Japan) region, close to one third of all IoT spending (hardware, software, services and connectivity combined) will be for the manufacturing industry in 2020. Also in other regions, manufacturing ranks first but with slower market shares of total IoT spending. In the US, for instance, IoT spend by the manufacturing industry will account for approximately 15 percent of total IoT purchases.

The manufacturing industry is leading in the IoT for various reasons: some are historical, others are related with the so-called next Industrial Revolution (Industry 4.0) and then there are the many uses cases and actual IoT deployments that offer rapid return and enable manufacturers to realize digital transformations from several perspectives: efficiency, automation, customer-centricity, competitive benefits and the advantages which are offered by using data across the manufacturing value chain and to tap into new revenue sources, a key aspect of digital transformation in manufacturing.[1]

2. IoT Use Cases in Manufacturing: Opportunities and Context

If we look at IoT in manufacturing from an IoT use case perspective (the areas where IoT is leveraged within a practical usage context), we see that one use case is sticking out: manufacturing operations, which is also the largest use case across all industries globally (the only region where manufacturing operations aren't the first use case is EMEA where freight monitoring comes first).

In 2016, manufacturing operations accounted for a total IoT spend of $102.5 billion (on the mentioned total of $178 billion), according to the same IDC 2017 release.

Manufacturing obviously covers many types of products, operations, processes and a vast space of activities, components, machines, people, partners, information systems and so forth. It is a long way from raw materials to finished goods and it is inevitably related with supply chains, logistics and transportation as well.

If we look at manufacturing as industrial production in any of the stages where raw materials are turned into products or products are used to build other products, it's clear that we see a huge market that is highly interconnected.[2]

It's among others in this sense that the Internet of Things almost by definition is key for the manufacturing industry in an integrated approach, further including technologies such as big data analytics, cloud, robotics and, most importantly perhaps, the integration of IT (information technology) and OT (operational technology).

In few other industries there are so many opportunities to leverage the IoT in connecting physical and digital, making various assets, such as machines, other production assets and the various object in a non-production sense, as well as a variety of product and manufacturing process parameters part of a vast information network. This is an important element as with manufacturing we typically tend to think about goods and products but the bigger opportunity for manufacturers lies in cyber-physical systems, a service economy model and the information opportunity.

3. IoT Use Cases: the Main Areas

The complexity and breadth of manufacturing shows in the numerous IoT use cases in manufacturing. Part of them are to be seen in the context of the digital/connected factory, others refer more to facility and asset management and still others relate to components such as safety/security and operations/logistics/ecosystems. Last but not least there is the customer preference and behavior part.

Below is a list with several IoT use cases in manufacturing and their benefits/realities.

(1) Manufacturing Operations

First comes manufacturing operations, as mentioned also the largest manufacturing IoT use case. In other words: operations of manufacturing facilities, involving several assets and operational personnel.

Manufacturing operations include the several elements which are typical in manufacturing operations management (MOM), such as asset management, intelligent manufacturing, performance optimization and monitoring, planning, human machine interaction, end-to-end operational visibility and these cyber-physical systems as we know them from Industry 4.0. In fact, cyber-physical systems and the IoT are twins in Industry 4.0. IoT spending in this vast use case (or even set of use cases) is good for over 57 percent of all IoT manufacturing investments.

(2) Production Asset Management and Maintenance

This is the second largest IoT use case in manufacturing and in reality also consists of a range of potential applications. It includes production asset monitoring and tracking, from location to the monitoring of parameters in several areas such as quality, performance, potential damage or breakdowns, bottlenecks, the list goes on. On top of performance and optimization, there is of course also the dimension of maintenance (as a result and/or in a predictive way). It's clear that asset management and maintenance in a manufacturing industry setting also go beyond pure production assets.

(3) Field Service

According to the mentioned IDC report, this is the third most important IoT use case in manufacturing. Here we leave the factory or manufacturing facility and go directly to an important area where manufacturers are service providers. From product-related services to business-related services: The (field) service organizations of manufacturers are important drivers of growth and, obviously, of margin. It's clear that information in the hyper-connected and hyper-aware digitized and IoT-enabled manufacturing ecosystem, along with the tools to plan, schedule and pro-actively service, are important differentiators.[3]

In an updated forecast on IoT spending in June 2017, IDC predicts that manufacturing will reach \$ 183 billion in 2017 from an IoT spending perspective. Production asset management spending would reach 45 billion.

(4) Other Manufacturing IoT Use Cases

Although the three mentioned areas account for the large majority of spending, there are several other manufacturing IoT use cases on levels of processes, assets and, let's not forget, people. Safety, security, worker protection (and productivity) and the many links between manufacturing in the strictest sense with connected services/operations/industries such as transport, supplier management and so forth all contribute to the vast IoT-related manufacturing solutions.

Vehicle and asset tracking, connected factory applications, staff safety applications, health monitoring (real-time), smart ventilation and air quality management, smart environmental measurement, access control (security), smart measurement of presence/levels of liquids, gases, radiation and dangerous materials (depending on the type of operation), asset protection, facility management, risk measurement, the list is long.

Last but not least there are the several related processes inside and outside the manufacturing facility. Production assets are one thing, activities such as preparing for shipment, packaging and quality control of manufactured goods another.

New Words and Expressions

customer-centricity　以客户为中心
leverage　n.　杠杆，杠杆作用，影响力
　　　　　v.　发挥杠杆作用，施加影响
stick out　突出，伸出手臂，伸展，(因为与众不同而)容易识别
EMEA　(Europe, the Middle East and Africa)欧洲、中东、非洲三地区的合称

last but not least　最后但同样重要的，最后但并非最不重要的
cyber-physical system　信息物理系统
IDC (International Data Corporation)　国际数据公司
ecosystem　n.　生态系统

Notes

[1] The manufacturing industry is leading in the IoT for various reasons: some are historical, others are related with the so-called next Industrial Revolution (Industry 4.0) and then there are the many uses cases and actual IoT deployments that offer rapid return and enable manufacturers to realize digital transformations from several perspectives: efficiency, automation, customer-centricity, competitive benefits and the advantages which are offered by using data across the manufacturing value chain and to tap into new revenue sources, a key aspect of digital transformation in manufacturing. 制造业是物联网的领军领域有多种原因：有些是历史性的，有些是与所谓的下一代工业革命（工业4.0）相关的，这其中有很多应用和实际的物联网部署，能够提供快速转变并使生产从几个方面实现数字革新：效率、自动化、用户为中心、竞争利润以及通过使用生产价值链的数据来挖掘新的收入来源带来的益处，而这也是制造中数字转换的关键点。

[2] If we look at manufacturing as industrial production in any of the stages where raw materials are turned into products or products are used to build other products, it's clear that we see a huge market that is highly interconnected. 如果我们将制造的任何一阶段看作是工业生产，其中原材料变成产品或者用于构造其他产品，那么我们就可以清楚地看到一个高度互联的巨大市场。

[3] It's clear that information in the hyper-connected and hyper-aware digitized and IoT-enabled manufacturing ecosystem, along with the tools to plan, schedule and pro-actively service, are important differentiators. 很清楚，在基于物联网的高度连接、高度数字化制造生态系统中的信息，以及用于安排策划和支持积极服务的工具是重要的区分特征。

Unit 8

A The Conversion of the Elements of a Sentence（句子成分的转换）

英汉两种语言，由于表达方式不尽相同，在具体英译汉时，有时往往需要转换一下句子成分，才能使译文逻辑正确、通顺流畅、重点突出。句子成分的转换作为翻译的一种技巧，其内容和形式都比较丰富，运用范围也相当广泛，共包括四个方面的内容：

1. 转译为汉语的主语

（1）宾语转译为主语

Example 1：Considerable use has been made of *these data*.

译文：这些数据得到了充分的利用。

Example 2：Semiconductors offer more *resistance* than good conductors.

译文：半导体的电阻大于良导体。

Example 3：Light beams can carry more *information* than radio signals.

译文：光束运载的信息比无线电信号运载的信息多。

Example 4：From an economic point of view, integrated circuits mean much lower *costs* through the use of automated mass production methods.

译文：从经济观点来看，通过采用自动化批量生产方法，集成电路的成本大大降低。

Example 5：Automatic lathes perform basically similar *functions* but appear in a variety of forms.

译文：各种自动车床的作用基本相同，但形式不同。

（2）谓语转译为主语

当英语中的谓语动词在译成汉语时名词化，便可转译为汉语的主语。

Example 1：Fuzzy control *acts* differently from conventional PID control.

译文：模糊控制的作用不同于传统的 PID 控制。

Example 2：Copper *conducts* electricity very well.

译文：铜的导电性能良好。

Example 3：In the stomach, this acid *functions* to kill bacteria in foods.

译文：在胃里，这种酸的作用是杀死食物中的细菌。

2. 转译为汉语的谓语

（1）定语转译为谓语

Example 1：This is a large amount of energy *wasted* due to the friction of commutator.

译文：换向器引起的摩擦损耗了大量的能量。

Example 2：An electronic computer is a kind of machine *giving accurate computation* at high speed.

译文：电子计算机是一种机器，它能高速地进行精确运算。

Example 3：In the forward direction the diode has a very *low* resistance.

译文：正向导通时二极管的电阻很低。

Example 4：Even well before, scientists recognized that light has a *vast* capacity for transmitting information.

译文：甚至很久以前，科学家们就认识到光传播信息的容量极大。

Example 5：Gear pumps operate on the very *simple* principle.

译文：齿轮泵的工作原理很简单。

（2）主语转译为谓语

当主语是动作名词，而谓语动词为系动词或被动形式时，主语就可以转译为汉语的谓语。

Example 1：*Care* should be taken to protect the instrument from dust and damp.

译文：应当注意保护仪器免受灰尘和潮湿。

Example 2：Then came the *development* of the integrated circuits.

译文：然后，集成电路发展起来了。

Example 3：The *statement* of the first law of motion is as follows.

译文：运动第一定律叙述如下。

（3）状语转译为谓语

Example 1：Television brings the world into your own home *in sight and sound*.

译文：电视使你在自己家里就能看到和听到世界上发生的事。

Example 2：Fission is accompanied by *the liberation of neutrons*.

译文：随着裂变释放出中子。

（4）表语转译为谓语

Example 1：Electronics is *the study* of the flow of electrons and the application of such knowledge to practical problems in communications and controls.

译文：电子学研究电子的运动规律，并把这种知识运用于通信和控制的实际问题中。

Example 2：In its early stages cancer is a local *collection* of cells and is easily removed.

译文：早期癌症的癌细胞集中在身体的某一部分，因而很容易切除。

3. 主语转译为汉语的宾语

Example 1：*Much progress* has been made in electrical engineering in less than a century.

译文：在不到一个世纪中，电气工程有了很大进展。

Example 2：*The resistance* can be determined provided that the voltage and current are know.

译文：只要知道电压和电流，就能求出电阻值。

4. 主语转译为汉语的定语

Example 1：*Transistors* are small in size, light in weight and consume very little power.

译文：晶体管的体积小，重量轻，消耗功率小。

Example 2：In fission process the *fission fragments* are very radioactive.

译文：在裂变过程中，裂变碎块的放射性很强。

Example 3：*Electronic circuits* work a thousand times more rapidly than nerve cells in the human brain.

译文：电子电路的运行速度比人脑中的神经细胞要快 1000 倍。

Exercises

Translate the following sentences into Chinese by the conversion techniques of the elements of a sentence. Make sure to translate the underlined words or phrases properly.

(1) The wings are responsible for keeping the airplane in the air.

(2) Machinery has made the products of manufactories very much cheaper than formerly.

(3) Radar works in very much the same way as the flashlight.

(4) Neutron has a mass slightly larger than that of proton.

(5) Scientists in that country are now supplied with necessary books, epuipment and assistant, that will ensure success in their scientific research.

(6) The result of his revolutionary design is that the engine is much smaller, works more smoothly, and has fewer moving parts.

(7) Industrial electronics started with transducers which allowed remote monitoring of processes, especially those which involved high temperatures or dangerous substances.

(8) These three colors, red, green, and violet, when combined, produced white.

(9) Where there is nothing in the path of beam of light, nothing is seen.

(10) Because he was convinced of the accuracy of this fact, he stuck to his opinion.

(11) The idea of obtaining potable water from wastewater is a psychologically difficult one for many people to accept.

(12) But in the case of electrical energy, the development from two completely chance observations was direct and very swift.

B Introduction to PID Controllers（比例积分微分控制入门）

Although the new and effective theories and design methodologies being continually developed in the automatic control field, proportional-integral-derivative (PID) controllers are still by far the most widely adopted controllers in industry owing to the advantageous cost/benefit ratio they are able to provide.[1] In fact, although they are relatively simple to use, they are able to provide a satisfactory performance in many process control tasks. Indeed, their long history and the know-how that has been devised over the years has consolidated their usage as a standard feedback controller. However, the availability of high-performance microprocessors and software tools and the increasing demand of higher product quality at reduced costs still stimulate researchers to devise new methodologies for the improvement of performance and/or for an easier use of them.[2] This is proven by the large number of publications on this topic (especially in recent years) and by the increasing number of products available on the market. Actually, much of the effort of researchers has been concentrated on the development of new tuning rules for the selection of the values of the PID parameters.

Although this is obviously a crucial issue, it is well-known that a key role in the achievement of high performance in practical conditions is also played by those functionalities that have to (or can) be added to the basic PID control law. Thus, in contrast to other articles on PID control, this article focuses on some of these additional functionalities and on other practical problems that a typical practitioner has to face when implementing a PID controller (for scalar linear systems). Recent advances as well as more standard methodologies are presented in this context.

A PID controller is a three-term controller that has a long history in the automatic control field, starting from the beginning of the 20th century.[3] Owing to its intuitiveness and its relative simplicity, in addition to satisfactory performance which it is able to provide with a wide range of processes, it has become in practice the standard controller in industrial settings.[4] A PID controller is a generic control loop feedback mechanism (controller) widely used in industrial control systems. A PID is the most commonly used feedback controller. A PID controller calculates an "error" value as the difference between a measured process variable and a desired setpoint. The controller attempts to minimize the error by adjusting the process control inputs.

The PID controller calculation (algorithm) involves three separate parameters, and is accordingly sometimes called three-term control: the proportional, the integral and derivative values, denoted P, I, and D. Heuristically, these values can be interpreted in terms of time: P depends on the *present* error, I on the accumulation of *past* errors, and D is a prediction of *future* errors, based on current rate of change. The weighted sum of these three actions is used to adjust the process via a control element such as the position of a control valve or the power supply of a heating element.

In the absence of knowledge of the underlying process, a PID controller is the best controller. By tuning the three constants in the PID controller algorithm, the controller can provide control action designed for specific process requirements. The response of the controller can be described in terms of the responsiveness of the controller to an error, the degree to which the controllerovershoots the setpoint and the degree of system oscillation. Note that the use of the PID algorithm for control does not guarantee optimal control of the system or system stability.

Some applications may require using only one or two modes to provide the appropriate system control. This is achieved by setting the gain of undesired control outputs to zero. A PID controller will be called a PI, PD, P or I controller in the absence of the respective control actions. PI controllers are fairly common, since derivative action is sensitive to measurement noise, whereas the absence of an integral value may prevent the system from reaching its target value due to the control action.

The PID control has been evolving along with the progress of the technology and nowadays it is very often implemented in digital form rather than with pneumatic or electrical components.[5] It can be found in virtually all kinds of control equipment, either as a stand-alone (single-station) controller or as a functional block in programmable logic controllers (PLCs) and distributed control systems (DCSs).[6] Actually, the new potentialities offered by the development of the digital technology and of the software packages has led to a significant growth of the research in the PID control field: New effective tools have been devised for the improvement of the analysis and design methods of the basic algorithm as well as for the improvement of the additional functionalities that are implemented with the basic algorithm in order to increase its performance and its ease of use. The success of the PID controllers is also enhanced by the fact that they often represent the fundamental component for more sophisticated control schemes that can be implemented when the basic control law is not sufficient to obtain the required performance or a more complicated control task is of concern.[7] The fundamental concepts of PID control are introduced with the aim of presenting the rationale of the control law. In particular, the meaning of the three actions is explained and the tuning issue is briefly discussed. The different forms for the implementation of a PID control law are also addressed.

New Words and Expressions

methodology	n.	方法论	oscillation n.	振荡
consolidate	v.	巩固	heuristically adv.	启发式地
algorithm	n.	算法	stand-alone n.	单机
intuitiveness	n.	直观		

Notes

[1] Although the new and effective theories and design methodologies being continually developed in the automatic control field, proportional-integral-derivative (PID) controllers are still by far the most widely adopted controllers in industry owing to the advantageous cost/benefit ratio they are able to provide. 尽管在自动控制领域不断有新的有效的理论和设计方法被创造出来，PID 至今仍然是工业中最广泛应用的控制器，这归功于它能提供具有优势的投入产出比。

[2] However, the availability of high-performance microprocessors and software tools and the increasing demand of higher product quality at reduced costs still stimulate researchers to devise new methodologies for the improvement of performance and/or for an easier use of them. 然而，高性能的微处理器和软件工具的使用，加上在减少成本的前提下对高生产质量不断增长的要求，激励着研究人员去设计新的方法来提高 PID 的性能和降低它们的操作难度。

[3] A PID controller is a three-term controller that has a long history in the automatic control field, starting from the beginning of the 20th century. 比例积分微分控制器（PID）是一种在自动控制领域拥有久远历史的三项控制器，它始创于 20 世纪初期。

[4] Owing to its intuitiveness and its relative simplicity, in addition to satisfactory performance which it is able to provide with a wide range of processes, it has become in practice the standard controller in industrial settings. 除了为广泛的工业过程提供满意的运行结果外，它的直观和相对简易性使它成为实际中工业环境的标准控制器。

[5] The PID controller has been evolving along with the progress of the technology and nowadays it is very often implemented in digital form rather than with pneumatic or electrical components. PID 控制器也随着科学技术的发展而不断进化，如今它更常以数字化形式实现而不是以气动或电子元件形式实现。

[6] It can be found in virtually all kinds of control equipment, either as a stand-alone (single-station) controller or as a functional block in programmable logic controllers (PLCs) and distributed control systems (DCSs). 无论是作为单机控制器还是作为可编程控制器（PLC）和分布

控制系统（DCS）中的函数模块，我们几乎能在所有的控制设备上发现它。

[7] The success of the PID controllers is also enhanced by the fact that they often represent the fundamental component for more sophisticated control schemes that can be implemented when the basic control law is not sufficient to obtain the required performance or a more complicated control task is of concern. 当基本控制律不足以达到需要的性能指标或者要处理一个更复杂的控制任务时，需要更复杂的控制方案。然而，PID 往往是其基本的元件，这一事实推动了 PID 在工业上的成功。

C Applications of Social Robots（社交机器人的应用）

Social robotics, an important branch of robotics, has recently drawn increasing attention in many disciplines, such as computer vision, artificial intelligence, and mechatronics, and has emerged as an interdisciplinary undertaking. While a number of social robots have been developed, a formal definition of social robot remains unclear and different practitioners have defined it from different perspectives. In Wikipedia, social robot is specified to be an autonomous robot that interacts and communicates with humans or other autonomous physical agents by following some social rules. While there are some differences among these definitions, a common characteristic can be reflected and we define a social robot as follows:

"A social robot is a robot which can execute designated tasks, and the necessary condition turning a robot into a social robot is the ability to interact with humans by adhering to certain social cues and rules."[1]

It is generally believed that human-robot interaction (HRI) is the heart of a social robot, and the interaction capability is the most important factor for a social robot. There have been several attempts on HRI of social robots in recent years, and two representative examples are Kismet (developed by MIT in 2002) and ASIMO (developed by the Honda company in 2003). Breazeal and colleagues in the MIT Media Lab developed Kismet, which has a human-like head to communicate with humans. Honda developed ASIMO which also demonstrated human-like characteristics to assist humans. More recently, a number of social robots have been designed and already or potentially applied to people's everyday lives as companions, assistants, and entertainment toys. For example, RoboX is a tour-guide robot at the Swiss National Exhibition Expo. 02, Sony AIBO has entertained humans to bring happiness, and Kismet has assisted people for social interaction studies. In this survey, we mainly focus on the interaction between a social robot and humans, also called HRI in social robots.

Generally, HRI of a social robot consists of three parts: perception, action and an "intermediate" mechanism. Perception refers to an environmental information acquisition and analysis module; action means the responses a robot made after it receives motor-control signals; and the intermediate mechanism is equivalent to "a robot's brain" connecting perception and action to produce motor-control signals according to the results from the perception analysis module.[2] In a general HRI framework in a social robot, the perceptual system dominates the whole HRI because it acts as a bridge between a social robot and the outside environment. Only the robot accurately understands the surrounding world, it can give meaningful responses to the correct subject. Due to high potentials and great importance of a perceptual system, we review state-of-the-art perception methods of HRI in social robots from three aspects: feature extraction, dimensionality reduction, and semantic understanding. Features refer to the information conveyed by raw signals, such as energy, color, and texture. Generally speaking, there are four classes of signals captured by a social robot: visual-based, audio-based, tactile-based, and range sensors-based. They are collected by cameras, microphones, tactile sensors, and laser range finders, respectively. Semantic understanding includes advanced analysis on these extracted features, such as sound localization, face detection, and emotion recognition.

While many challenges are encountered when applying social robots in real-world applications, there are still some social robots developed to assist our daily life, such as Kismet, iCub, and Robovie, as well as some commercially available social robots including Sony's AIBO and NEC's PaPeRo.[3] In this section, we introduce some representative social robots and show how the perception methods work in these robots.

1. Robots as Test Subjects

(1) Social Development

For the research of social development, infants and young children are the popular studied subjects. Hence, social robots as test subjects are generally designed as infant-like or child-like, which have some learning skills such as imitation and joint attention.

Cog is a representative robot, which is developed for human cognition and developmental psychology. It has 22 DOFs (degrees of freedom) distributed on the arms, torso, neck, and eyes. This robot can manipulate objects, show head postures, and move eyes. To achieve these goals, visual signals were adopted for face detection and object segmentation.

iCub is another representative social robot, and its objective is to offer a platform for cognition investigation. It comes from RoboCub, a 5-year project funded by the European Commission through Unit E5 "Cognitive Systems, Interaction & Robotics". iCub is particularly designed as a 3-4 years old child. It can interact with the outside environments with the head, neck, arms, torso, and legs. For example, iCub can follow objects by orientating its head and eyes, dexterously manipulate objects with its hands, crawl and sit up like a child. Human detection, object recognition, and sound localization were implemented in this robot, respectively.

(2) Social Interaction

An important way to learn new knowledge from environments is social interactions. To investigate social interactions between humans, a natural approach is to design human-like robots. According to the pre-defined personality, robots make reasonable responses through speech or body languages when interacting with humans.

Kisme, a well-known robot head, is designed to investigate social interactions between caregivers and babies. It interacts with humans through facial expression, body gesture, and vocal babbles. Its perception abilities include object detection, object recognition, and emotion recognition. Kismet has been a popular prototype in many research groups to develop their social robots.

Besides human-like appearances, MIT Media Lab in collaboration with Stan Winston Studio and

DARPA has developed a robot named Leonardo with animatronic characteristics. Leonardo has a highly mobile face and body, and can learn new skills and cooperate with users by social interactions. It is able to remember the users, execute the verbal commands, and share attentions. There are several perception abilities on this robot, such as face recognition, speech recognition, object recognition and tracking, and tactile perception.

2. Socially Assistive Robotics

There are some socially assistive robots which provide services to humans in public places or domestic environments. Since it is closely related to people's lives, more recognition abilities and social cues are required in these robots. Moreover, due to the difference between the public places and the domestic environments, the perception tasks are also distinct.

(1) Robots in Public Places

Robots in pubic places refer to the robots used in museums, supermarkets, shopping centers, and childhood education centers. Different places require different tasks. For example, if the robot is applied in museums, it is used to introduce the exhibits to humans; when it is applied for child-hood education, it is used to help teacher to organize or teach children.

GRACE is a representative robot which is used in public places. It firstly appeared in AAAI 2002 to perform the Robot Challenge, where the tasks were to navigate the registration desk of the conference center, come to the conference room, and give a presentation.[4] In AAAI 2005, GRACE's task was extended to find a person. In both AAAI 2002 and 2005, GRACE successfully completed speech recognition and human tracking in real-world environments.

RoboX, a tour-guide robot, appeared at the Swiss National Exhibition Expo. 02, and Rackham, another tour-guide robot, appeared in Mission BioSpace exhibition. Their task is to present exhibits for tourists. To implement such task, they need to navigate environments, introduce themselves, ask visitors to choose a visiting destination, and give the corresponding presentations. The perception tasks include motion detection and tracking, face detection and recognition, speech recognition, and gesture classification.

Robovie is a robot with different functions. Robovie II aims to help the elders shop in the Apita-Seikadai supermarket in Kyoto, and Robovie III aims to provide directions for people and invite customers to visit the shop. Robovie II includes face and gesture recognition to recognize human faces and hand poses of the users. Robovie III contains human behavior analysis to identify the users' walking styles, and directions.

RUBI is a three-feet tall robot. It consists of a head, two arms and touch screen, and is designed to assist teachers for early childhood education. RUBI was set at the Early Childhood Education Center at the University of California, San Diego, interacting with the children with 18-24 months old. It can teach children numbers, colors and some basic concepts, and schedule proper lessons and assist teachers according to the children's emotional responses. It contains some perception functions such as face detection and tracking, and emotion recognition.

(2) Robots in Domestic Environments

Robots in domestic environments mean the robots used at home. Christensen has summarized three domestic environments: entertainment, everyday tasks, and assistance to the elder and handicapped. Lohse and colleagues have proposed a functionality-based categorization including health care, companionship, entertainment, toy, pet, and personal assistants. Different from the robots which are used in various and different public places, the robots in domestic environments are generally employed at home. Hence, their users are prespecified.

ARMAR III, developed in University of Karlsruhe, can execute tasks in household environments. The robot can open and close a dishwasher, pick up cups and dishware, place them anywhere within reach, and plug an electrical appliance into the wall in a kitchen. Moreover, it can interact with humans by using speech and gesture recognition. To achieve these goals, ARMAR III has a head, two arms, two hands, a torso and a mobile platform, to perform human verification and tracking, head pose estimation, and sound classification.

PaPeRo is a social robot designed by the NEC Corporation and has been commercially available. It is a personal robot which can care for children and provide assistance to elders. Several applications were developed: speech conversation, face recognition, touching reaction, roll-call and quiz game designing, communications through phone or PC, learning greetings, and story telling.[5] Moreover, speakers and LEDs are used to produce speech and songs and display PaPeRo's internal status, respectively.

Huggable was developed by the personal robots research group from the MIT Media Lab. Different from the above described robots, it uses tactile-based signals for healthcare, education, and social communication applications. Huggable has the appearance of Teddy bear, and is covered with a full-body sensitive skin containing more than 1,500 sensors. Hence, it can detect and recognize pressure from the outside world. In addition, cameras and micro phones are used to sense the surrounding environment. After semantically analyzing the collected data, the robot can convey a personality—rich character through some gestures and expressions. Moreover, it can be remotely controlled and applied to monitor the elder and children through a web interface.

3. Robots for Studying Human-Robot Interaction

HRI plays an important role on the above robots even if they are developed for different applications. HRI can help a robot to learn new knowledge, obtain useful information, and attract persons. Motivated by these significant roles, more and more researchers have focused on HRI in social robots and developed several social robots, such as MEXI, ROMAN, BARTHOC, and Fritz for HRI studies.

MEXI is a robot head to detect faces and recognize emotions. It responds to faces and emotions with different facial expressions such as smiling, sulking, and looking around. Additionally, it uses a commercially available software to respond with speech. Similar to MEXI, ROMAN is another humanoid robot head. ROMAN can follow the detected human using its eyes and neck, and non-verbally communicate with humans using facial expressions.

BARTHOC and BARTHOC Jr. are two upper-body humanoid robots for studying HRI. The main differences between them include the sizes and the weights. BARTHOC mimics adult person, and BARTHOC Jr. imitates a four-year old child. Both of them can perform emotion recognition, voice detection and human tracking. BIRON is another robot developed by Bielefeld University, and has ported its several basic functions to BARTHOC such as human tracking. Since the appearances of BARTHOC and BARTHOC Jr. are more human-like than that of BIRON, speech-based emotion recognition has been developed and applied in BARTHOC Jr. [6] With the recognized results, the robot can mirror human's affective states by its own facial images.

Fritz is another body-based social robot developed for HRI. It was originally developed to play soccer, and is now a platform to study multimodal communication with humans. It has perception abilities such as face detection, face tracking, and speaker localization. By using body gestures, facial expressions and synthesized speech, the robot can attract person's interest on communication.

To allow robot researchers to focus on the specific fields, some companies and research institutions have developed the robot platforms from both hardware and software aspects. These robot platforms can implement several basic functions. For example, HOAP series from Fujitsu, QRIO from Sony, HUBO (KHR-3) from KAIST, and HRP series from Kawada Industries are humanoid robot platforms. While they are of different appearances and sizes, their basic functions are similar. These robots have perception abilities like object recognition and tracking, sound localization and recognition. QRIO can also recognize the users' voice and face. Another two robot platforms are PR2 from Willow Garage and YouBot from KUDA. They have been commercially available now. PR2 is designed as a personal robot with mobility. It can navigate human environments and manipulate some objects in the environments. For YouBot, it consists of a manipulator and a mobile base. Hence, it has mobility and can grasp some objects. Based on these developed robot platforms, several perception tasks such as emotion recognition and gesture recognition can be included for HRI study. [7]

4. Discussion

In this section, we reviewed several representative social robots according to their different applications including robots as test subjects, robots providing social assistance, and robots for studying HRI. We mainly focused on the employed perception methods on HRI of these robots. [8] As Goodrich and Schultz defined: "Human-Robot Interaction (HRI) is a field of study dedicated to understanding, designing, and evaluating robotic systems for use by or with humans." A significant challenge of the study is how to achieve natural communication between humans and robots. To address this, one primary and key factor is using different perception methods in HRI.

As we described above, different social robot applications affect the design of perception systems. For the robots as test subjects and for studying HRI, they are usually used in lab environments which are much simpler than the real-world environments, and the perception tasks mainly include human detection and tracking, face detection, recognition, and tracking, gesture recognition, sound localization and recognition, and emotion recognition. [9] In addition, when humans use objects such as toys to interact with the robots, the robots should be able to detect and track the objects. If humans want to have physical contacts with the robots, the robots should also have the ability of tactile detection and classification. For these perception tasks, visual signals acquired by cameras, audio signals acquired by microphones, and tactile signals acquired by tactile sensors are usually employed. Since the number of users under such scenarios is limited, the environment is comparatively simple, and the robots normally have no mobility, the extracted features and semantic understanding methods are usually simple.

For socially assistive robotics, since they could be used in home, offices, or hospitals, the extracted features and semantic understanding methods are normally different from those used in robots as test subjects and for studying HRI. For example, if the robot is applied in a shopping center, its human tracking system usually requires to simultaneously track several persons rather than only one user. Usually, due to the complex application environments like uncontrolled illumination and background, it will need more robust features and semantic understanding methods to help the robot successfully fulfill the tasks. In addition, because of the mobility of several socially assistive robotics, besides visual, audio, and tactile signals, laser reading acquired by range finders is another useful modality to interact with the outside environment for the robots.

Generally speaking, the selection of sensors and feature extraction methods depend on the semantic understanding tasks, and semantic understanding tasks depend heavily on the application scenarios of robots. For instance, the semantic understanding tasks for the robot used in a museum and that used in a childhood education center are different due to different objectives. Correspondingly, the sensors and feature extraction methods are also different for distinct semantic understanding tasks. Four basic rules are applied to select the sensors and extract features in social robots: high stability, fast speed, high accuracy, and high autonomy.

New Words and Expressions

mechatronics *n.* 机电一体化		infant *n.* 婴儿
interdisciplinary *adj.* 跨学科的		DOFs (degrees of freedom) 自由度
semi-autonomous *adj.* 半自主的		torso *n.* 躯干
HRI (human-robot interaction) 人机交互		dexterously *adv.* 巧妙地
potentially *adv.* 潜在地		vocal babbles 牙牙学语
perception *n.* 知觉		prototype *n.* 原型,雏形
perceptual *adj.* 感性的		navigate *v.* 驾驶,续航
dimensionality reduction 降维		handicapped *adj.* 残疾的,有生理缺陷的
semantic *adj.* 语义的		pre-specified *adj.* 预先规定的
tactile-based *adj.* 基于触觉的		verification *n.* 核实,证明
segmentation *n.* 分割		sulk *v.* 生闷气
enumerate *v.* 列举		

Notes

[1] A social robot is a robot which can execute designated tasks, and the necessary condition turning a robot into a social robot is the ability to interact with humans by adhering to certain social cues and rules. 社交机器人是一种能够执行特定任务的机器人,一个机器人成为社交机器人的必要条件就是能够遵循一些社交提示和规则与人类交流互动。

[2] Perception refers to an environmental information acquisition and analysis module; action

means the responses a robot made after it receives motor-control signals; and the intermediate mechanism is equivalent to "a robot's brain" connecting perception and action to produce motor-control signals according to the results from the perception analysis module. 知觉是指环境信息采集与分析模块；行动是指机器人接收电动机控制信号后做出的响应；中间机构相当于"机器人的大脑"，它连接着感知和行动，并根据感知分析模块的结果产生运动控制信号。

[3] While many challenges are encountered when applying social robots in real-world applications, there are still some social robots developed to assist our daily life, such as Kismet, iCub, and Robovie, as well as some commercially available social robots including Sony's AIBO and NEC's PaPeRo. 社交机器人在实际应用中虽然遇到许多挑战，但仍然研发出来一些社交机器人来帮助我们的日常生活，如 Kismet、iCub 和 Robovie，以及一些商业销售的社交机器人，包括索尼的 AIBO 和 NEC 的 PaPeRo。

[4] It firstly appeared in AAAI 2002 to perform the Robot Challenge, where the tasks were to navigate the registration desk of the conference center, come to the conference room, and give a presentation. 它首先在 2002 年出现在美国智能协会的机器人挑战赛上，任务是浏览会议中心的登记处，并来到会议室进行演讲。

[5] Several applications were developed: speech conversation, face recognition, touching reaction, roll-call and quiz game designing, communications through phone or PC, learning greetings, and story telling. 如下几个应用程序已经被开发：语音对话、人脸识别、触摸反应、游戏设计的点名提问、通过手机或计算机通信、学习打招呼和讲故事。

[6] Since the appearances of BARTHOC and BARTHOC Jr. are more human-like than that of BIRON, speech-based emotion recognition has been developed and applied in BARTHOC Jr. 因为 BARTHOC 和 BARTHOC Jr. 的出场比 BIRON 更像人类，基于语音的情感识别已经开发并应用在 BARTHOC Jr. 上。

[7] Based on these developed robot platforms, several perception tasks such as emotion recognition and gesture recognition can be included for HRI study. 基于这些开发机器人的平台，如情感识别和手势识别这些直觉任务可以包括在人机交互研究中。

[8] In this section, we reviewed several representative social robots according to their different applications including robots as test subjects, robots providing social assistance, and robots for studying HRI. 在这部分中，我们根据不同的应用回顾了几个有代表性的社交机器人，包括作为试验对象的机器人、提供社会援助的机器人，还有研究人机交互的机器人。

[9] For the robots as test subjects and for studying HRI, they are usually used in lab environments which are much simpler than the realworld environments, and the perception tasks mainly include human detection and tracking, face detection, recognition and tracking, gesture recognition, sound localization and recognition, and emotion recognition. 对于用来实验研究人机交互的机器人，它们通常被用在比现实世界简单得多的实验室环境中，感知任务主要包括人检测和跟踪、人脸检测、识别跟踪、手势识别、声源定位识别以及情感识别。

PART I　Techniques of EST E-C Translation（科技英语英译汉翻译技巧）

Unit 9

A　Translation of Passive Sentences（被动句的翻译）

在有些特定的语境下，特别是在科技文体中，客观性的表达非常重要，因为科技英语描述的主体往往是客观的事物、现象或过程，而客体往往是从事某项工作的人或装置，这时使用被动语态不仅比较客观，而且还可以使读者的注意力集中在描述的事物、现象或过程，即主体上。因此，科技英语中广泛使用被动语态，其出现的缘由多种多样，从语言应用方面考虑，主要有以下几个方面：

（1）不知道行为者是谁或不必说明行为者是谁

Example 1：The book has already been translated into many languages.

译文：该书（不知作者是谁或不必说明行为者是谁）已译成好几种语言。

（2）对行为对象比对行为者更感兴趣

Example 2：The work must be finished at once.

译文：该项工作必须马上完成。（谁来完成不是重要的，重要的是工作必须完成）

（3）为了某种缘故不愿说出行为者是谁

Example 3：It is generally considered not advisable to act that way.

译文：大家（不愿或不便说出具体是谁）认为这样做不妥。

（4）为了便于连贯上下文及语句安排上的方便

Example 4：They are going to build a library here next year. It is going to be built beside the classroom building.

译文：他们准备明年在这里建一座图书馆，就建在教学楼旁。

与英语不同，汉语中的被动往往有"受制于人"的感觉，使人感到不习惯、不舒服。所以，若不是特别强调叙述对象，一般都应将英语中的被动语态非被动化。通常英语中的被动句有以下几种翻译方法：

1. 翻译成汉语的被动句

对于具有"$n.$ + be + 动词的过去分词 + by + $n.$"结构的被动句，为了强调动作，或不知道动作发出者，或在特定的上下文里，为了使前后分句的主语保持一致，重点突出，语义连贯，语气流畅，可将英语的被动句译为汉语被动句。在表达上除了采用"被"字外，还可以根据汉语搭配的需要选用"遭、经过、给、加以、用……来、为……所、使、由……、受到"等词。

Example 1：The machine tools are controlled by PLC.

译文：机床由可编程控制器控制。

Example 2：The price of computer can be cut down to ten percent by the competition among suppliers.

译文：供应商之间的竞争能使计算机的价格下降10%。

Example 3：The strength of an electric current can be measured by the rate at which it flows through a wire.

译文：电流强度可以由电流流过导线的速率来测定。

Example 4：Other questions will be discussed briefly.

译文：其他问题将简单地加以讨论。

2. 翻译成带表语的主动句

对于具有 "$n.$ + be + 动词的过去分词" 结构的被动句可以翻译成带表语的主动句。

Example 1：This kind of device is much needed in the speed-regulating system.

译文：这种装置在调速系统中是很重要的。

Example 2：The energy of motion is called kinetic energy.

译文：运动的能量称为动能。

Example 3：The volume is not measured in square millimeters. It is measured in cubic millimeters.

译文：体积不是以平方毫米计量的。它是以立方毫米计量的。

Example 4：The decision to attack was not taken lightly.

译文：进攻的决定不是轻易做出的。

3. 将主语翻译成宾语

Example 1：The quartz crystal does not vibrate at certain frequency until the voltage is applied.

译文：直到电压加上去后，石英晶体才会以某一频率振荡。

Example 2：Since numerical control was adopted at machine tools, the productivity has been raised greatly.

译文：自从机床采用数控以来，生产率大大提高了。

Example 3：When power is spoken of, time is taken into account.

译文：说到功率时，总是把时间计算在内。

Example 4：Friction can be reduced and the life of the machine can be prolonged by lubrication.

译文：润滑能减少摩擦，延长机器寿命。

Example 5：A right kind of fuel is needed for an atomic reactor.

译文：原子反应堆需要一种合适的燃料。

Example 6：By the end of the war, 800 people had been saved by the organization, but at a cost of 200 Belgian and French lives.

译文：大战结束时，这个组织拯救了800人，但那是以200多比利时人和法国人的生命为代价的。

4. 句子中增加主语

当被动句中有形式主语或具有泛指的含义时，可以加入"人们""大家""有人"等，译成主动句。

Example 1：A few years ago it was thought unbelievable that the computer could have so high speed as well so small volume.

PART I Techniques of EST E-C Translation（科技英语英译汉翻译技巧）

译文：几年以前人们认为计算机能具有如此高的运行速度和如此小的体积是一件难以置信的事。

Example 2：Fuzzy control is found an effective way to control the systems without precise mathematic models.

译文：人们发现模糊控制是一种控制不具备精确数学模型系统的有效方法。

Example 3：Damage from the increased use of chemicals has been felt.

译文：我们已经觉察到了增加使用化学物质带来的危害。

Example 4：If one or more electrons are removed, the atom is said to be positively charged.

译文：如果原子失去了一个或多个电子，我们就说该原子带正电荷。

Example 5：It could be argued that the radio performs this service as well, but on television everything is much more living, much more real.

译文：可能有人会指出，无线电广播同样也能做到这一点，但还是电视屏幕上的节目要生动、真实得多。

Example 6：It is imagined by many that the operations of the common mind can be by no means compared with these of the scientists, and that they have to be acquired by a sort of special training.

译文：许多人认为，普通人的思维活动根本无法与科学家的思维过程相比，而且认为这些思维过程必须经过某种专门的训练才能掌握。

另外，下列的结构也可以通过这一方法翻译：

It is asserted that... 有人主张……

It is believed that... 有人认为……

It is generally considered that... 大家（一般人）认为……

It is well known that... 大家知道（众所周知）……

It will be said... 有人会说……

It was told that... 有人曾经说……

5. 翻译成无主语句

科技英语中的被动句有时也可以直接翻译成无主句，使句子结构更加紧凑。

Example 1：What kind of device is needed to make the control system simple?

译文：需要什么装置使控制系统简化？

Example 2：Some measures must be taken to control the water pollution.

译文：必须采取某些措施来控制水污染。

Example 3：Methods are found to take these materials out of the rubbish and use them again.

译文：现在已经找到了从垃圾中提取这些材料并加以利用的方法。

Example 4：New source of energy must be found, and this will take time.

译文：必须找到新的能源，这需要时间。

Example 5：Many strange new means of transport have been developed in our century, the strangest of them being perhaps the hovercraft.

译文：在我们这个世纪内研制了许多新奇的交通工具，其中最奇特的也许就是气垫船了。

另外,下列结构也可以通过这一方法翻译:
It is hoped that... 希望……
It is reported that... 据报道……
It is said that... 据说……
It is supposed that... 据推测……
It may be said without fear of exaggeration that... 可以毫不夸张地说……
It must be admitted that... 必须承认……
It must be pointed out that... 必须指出……
It will be seen from this that... 由此可见……

Exercises

Put the following sentences into Chinese by the translation techniques of passive sentences.

(1) Each of the body systems is regulated in some way by some part of the endocrine system.

(2) The supply of oil can be shut off unexpectedly at any time, and in any case, the oil wells will all run dry in thirty years or so at the present rate of use.

(3) Early fires on the earth were certainly caused by nature, not by Man.

(4) These signals are produced by colliding stars or nuclear reactions in outer space.

(5) Natural light or "white" light is actually made up of many colours.

(6) The behaviour of a fluid flowing through a pipe is affected by a number of factors, including the viscosity of the fluid and the speed at which it is pumped.

(7) The earth is hit from time to time by streams of electrically charged particles poured out by the sun.

(8) Over the years, tools and technology themselves as a source of fundamental innovation have largely been ignored by historians and philosophers of science.

(9) In other words mineral substances which are found on the earth must be extracted by digging, boring holes, artificial explosions, or similar operations which make them available to us.

(10) Nuclear power's danger to health, safety, and even life itself can be summed up in one word: radiation.

(11) Light rail systems are almost universally operated by electricity delivered through overhead lines.

(12) It is generally accepted that the experiences of the child in his first years largely determine his character and later personality.

B Electric Motors (电动机)

An electric motor converts electrical energy into mechanical energy. Most electric motors operate through interacting magnetic fields and current-carrying conductors to generate force, although electrostatic motors use electrostatic forces. The reverse process, producing electrical energy from mechanical energy, is done by generators such as an alternator or a dynamo. Many types of electric motors can be run as generators, and vice versa. For example a starter/generator for a gas turbine, or traction motors used on vehicles, often perform both tasks. Electric motors and generators are commonly referred to as electric machines.

Electric motors are found in applications as diverse as industrial fans, blowers and pumps, machine tools, household appliances, power tools, and disk drives. They may be powered by direct current (e. g. , a battery powered portable device or motor vehicle), or by alternating current from a central electrical distribution grid. The smallest motors may be found in electric wristwatches. Medium-size motors of highly standardized dimensions and characteristics provide convenient mechanical power for industrial uses. The very largest electric motors are used for propulsion of ships, pipeline compressors, and water pumps with ratings in the millions of watts. Electric motors may be classified by the source of electric power, by their internal construction, by their application, or by the type of motion they give.

The physical principle of production of mechanical force by the interactions of an electric current and a magnetic field was known as early as 1821. Electric motors of increasing efficiency were constructed throughout the 19th century, but commercial exploitation of electric motors on a large scale required efficient electrical generators and electrical distribution networks.

Some devices, such asmagnetic solenoids and loudspeakers, although they generate some mechanical power, are not generally referred to as electric motors, and are usually termed actuators and transducers, respectively.

The excitation to the field is dependent on the connections of the field winding relative to the armature winding. A number of choices open up, and they are treated briefly in the following sections.

1. Separately-excited DC Machines

If the field winding is physically and electrically separate from the armature winding, then the machine is known as a separately-excited DC (direct current) machine. The independent control of field current and armature current endows simple but high performance control on this machine, because the torque and flux can be independently and precisely controlled.[1] The field flux is controlled only by the control of the field current. Assume that the field is constant. Then the torque is proportional only to armature current, and, hence, by controlling only this variable, the dynamics the motor drive system can be controlled. With the independence of the torque and flux-production channels in this machine, it must be noted that it is easy to generate varying torques for a given speed and hence make torque generation independent of the operating speed. This is an important operating feature in a machine: the speed regulation can be zero. The fact that such a feature comes only with feedback control and not simply on open-loop operation is to be recognized.

2. Shunt-Excited DC Machine

If the field winding is connected in parallel to the armature winding, then the machine goes by the name of shunt-excited DC machine or simply by DC shunt machine. In this machine, the field winding does not need a separate power supply, as it does in the case of the separately-excited DC machine. For a constant input voltage, the field current and hence the field flux are constants in this machine. While it is good for a constant-input-voltage operation, it has troubling consequences for variable-voltage operation. In variable-input-voltage operation, an independent control of armature and field currents is lost, leading to a coupling of the flux and torque production channels in the machine.[2] This is in contrast to the control simplicity of the separately-excited DC machine. For a fixed DC input voltage, as torque is increased, the armature current increases, and hence the armature voltage drop also increases, while, at the same time, the induced EMF (Electromotive Force) is decreased. The reduction in the induced EMF is reflected in a lower speed, since the field current is constant in the machine. The drop in speed from its no-load speed is relatively small, and, because of this, the DC machine is considered constant-speed machine. Such a feature makes it unsuitable for variable-speed application.

3. Series-Excited DC Machine

If the field winding is connected in series with the armature winding, then it is known as the series-excited DC machine or DC series machine. It has the same disadvantage as the shunt machine, in that there is no independence between the control of the field and the armature currents. The electromagnetic torque of the machine is proportional to the square of the armature current, because the field current is equal to the armature current. At low speeds, a high armature current is feasible, with a large difference between a fixed applied voltage a small induced EMF. This results in high torque at starting and low speeds, making it an ideal choice for applications requiring high starting torques, such as in propulsion. With the dependence of the torque on the square of the armature current and the fact that the armature current availability goes down with increasing speed, torque-vs.-speed characteristic resembles a hyperbola.[3] Note that, at zero speed and low speeds, the torque is large but somewhat curtailed from the square current law because of the saturation of the flux path with high currents.

4. DC Compound Machine

Combining the best feature of the series and shunt DC machines by having both a series and shunt field in a machine leads to the DC compound machine. The manner in which the shunt-field winding is connected in relation to the armature and series field provides two kinds of compound DC machine.[4] If the shunt field encompasses the series field and armature windings, then that configuration is known as a long-shunt compound DC machine. The short-shunt compound DC machine has the shunt field in parallel to the armature winding. In the latter configuration, the shunt-field excitation is a slave to the induced EMF and hence the rotor speed, provided that the voltage drop across the armature resistance is negligible compared to the induced EMF.[5] Whether the field fluxes of the shunt and series field are opposing or strengthening each other gives two other configurations, known as differentially and cumulatively compound DC machines, respectively, for each of the long and short shunt connections.

New Words and Expressions

separately-excited DC machine 他励直流电机
shunt-excited DC machine 并励直流电机
series-excited DC machine 串励直流电机
electromagnetic *adj.* 电磁(体)的

pump *n.* 泵	漆的
dimension *n.* 尺寸，大小，维数	propulsion *n.* 推进，推进力
electrostatic *adj.* 静电（学）的，静电喷	solenoid *n.* 螺线管，圆筒形线圈

Notes

［1］ The independent control of field current and armature current endows simple but high performance control on this machine, because the torque and flux can be independently and precisely controlled. 磁场电流和电枢电流的独立控制使得电机具有简单却出色的控制性能，因为它可以分别精确地控制力矩和磁场的大小。

［2］ In variable-input-voltage operation, an independent control of armature and field currents is lost, leading to a coupling of the flux and torque production channels in the machine. 在可变电压输入的情况下，电枢电流和磁场电流的独立控制已不复存在，进而导致电机中励磁通路和力矩通路的耦合。

［3］ With the dependence of the torque on the square of the armature current and the fact that the armature current availability goes down with increasing speed, torque-vs.-speed characteristic resembles a hyperbola. 由于力矩和电枢电流的平方关系，以及电枢电流随着速度的上升而下降的趋势，整个力矩-转速特性曲线呈双曲线。

［4］ The manner in which the shunt-field winding is connected in relation to the armature and series field provides two kinds of compound DC machine. 并励磁场绕组和电枢以及串励磁场的连接方式提供了两种直流复励电动机。

［5］ In the latter configuration, the shunt-field excitation is a slave to the induced EMF and hence the rotor speed, provided that the voltage drop across the armature resistance is negligible compared to the induced EMF. 在后一种结构中，因为电枢电阻两端的电压和感应电动势相比可以忽略不计，所以并励磁场由感应电动势决定，进而也由转速决定。

C Data-Mining Concepts（数据挖掘概念）

Modern science and engineering are based on using first-principle models to describe physical, biological, and social systems. Such an approach starts with a basic scientific model, such as Newton's laws of motion or Maxwell's equations in electromagnetism, and then builds upon them various applications in mechanical engineering or electrical engineering. In this approach, experimental data are used to verify the underlying first-principle models and to estimate some of the parameters that are difficult or sometimes impossible to measure directly. However, in many domains the underlying first principles are unknown, or the systems under study are too complex to be mathematically formalized. With the growing use of computers, there is a great amount of data being generated by such systems. In the absence of first-principle models, such readily available data can be used to derive models by estimating useful relationships between a system's variables (i.e., unknown input-output dependencies).[1] Thus there is currently a paradigm shift from classical modeling and analyses based on first principles to developing models and the corresponding analyses directly from data.

We have gradually grown accustomed to the fact that there are tremendous volumes of data filling our computers, networks, and lives. Government agencies, scientific institutions, and businesses have all dedicated enormous resources to collecting and storing data. In reality, only a small amount of these data will ever be used because, in many cases, the volumes are simply too large to manage, or the data structures themselves are too complicated to be analyzed effectively. How could this happen? The primary reason is that the original effort to create a data set is often focused on issues such as storage efficiency; it does not include a plan for how the data will eventually be used and analyzed.

The need to understand large, complex, information-rich data sets is common to virtually all fields of business, science, and engineering. In the business world, corporate and customer data are becoming recognized as a strategic asset. The ability to extract useful knowledge hidden in these data and to act on that knowledge is becoming increasingly important in today's competitive world. The entire process of applying a computer-based methodology, including new techniques, for discovering knowledge from data is called data mining.

Data mining is an iterative process within which progress is defined by discovery, through either automatic or manual methods. Data mining is most useful in an exploratory analysis scenario in which there are no predetermined notions about what will constitute an "interesting" outcome.[2] Data mining is the search for new, valuable, and nontrivial information in large volumes of data. It is a cooperative effort of humans and computers. Best results are achieved by balancing the knowledge of human experts in describing problems and goals with the search capabilities of computers.

In practice, the two primary goals of data mining tend to be prediction and description. Prediction involves using some variables or fields in the data set to predict unknown or future values of other variables of interest. Description, on the other hand, focuses on finding patterns describing the data that can be interpreted by humans. Therefore, it is possible to put data-mining activities into one of two categories:

● predictive data mining, which produces the model of the system described by the given data set;

● descriptive data mining, which produces new, nontrivial information based on the available data set.

On the predictive end of the spectrum, the goal of data mining is to produce a model, expressed as an executable code, which can be used to perform classification, prediction, estimation, or other similar tasks. On the descriptive end of the spectrum, the goal is to gain an understanding of the analyzed system by uncovering patterns and relationships in large data sets. The relative importance of prediction and description for particular data-mining applications can vary considerably. The goals of prediction and description are achieved by using data-mining techniques, for the following primary data-mining tasks:

1) Classification. Discovery of a predictive learning function that classifies a data item into one of several predefined classes.

2) Regression. Discovery of a predictive learning function that maps a data item to a real-value

prediction variable.

3) Clustering. A common descriptive task in which one seeks to identify a finite set of categories or clusters to describe the data.

4) Summarization. An additional descriptive task that involves methods for finding a compact description for a set (or subset) of data.

5) Dependency modeling. Finding a local model that describes significant dependencies between variables or between the values of a feature in a data set or in a part of a data set.

6) Change and deviation detection. Discovering the most significant changes in the data set.

Current introductory classifications and definitions are given here only to give the reader a feeling of the wide spectrum of problems and tasks that may be solved using data-mining technology. [3]

The success of a data-mining engagement depends largely on the amount of energy, knowledge, and creativity that the designer puts into it. In essence, data mining is like solving a puzzle. The individual pieces of the puzzle are not complex structures in and of themselves. Taken as a collective whole, however, they can constitute very elaborate systems. As you try to unravel these systems, you will probably get frustrated, start forcing parts together, and generally become annoyed at the entire process, but once you know how to work with the pieces, you realize that it was not really that hard in the first place. The same analogy can be applied to data mining. In the beginning, the designers of the data-mining process probably did not know much about the data sources; if they did, they would most likely not be interested in performing data mining. Individually, the data seem simple, complete, and explainable. But collectively, they take on a whole new appearance that is intimidating and difficult to comprehend, like the puzzle. Therefore, being an analyst and designer in a data-mining process requires, besides thorough professional knowledge, creative thinking and a willingness to see problems in a different light.

Data mining is one of the fastest growing fields in the computer industry. Once a small interest area within computer science and statistics, it has quickly expanded into a field of its own. One of the greatest strengths of data mining is reflected in its wide range of methodologies and techniques that can be applied to a host of problem sets. Since data mining is a natural activity to be performed on large data sets, one of the largest target markets is the entire data-warehousing, data-mart, and decision-support community, encompassing professionals from such industries as retail, manufacturing, telecommunications, health care, insurance, and transportation. [4] In the business community, data mining can be used to discover new purchasing trends, plan investment strategies, and detect unauthorized expenditures in the accounting system. It can improve marketing campaigns and the outcomes can be used to provide customers with more focused support and attention. Data-mining techniques can be applied to problems of business process reengineering, in which the goal is to understand interactions and relationships among business practices and organizations.

Many law enforcement and special investigative units, whose mission is to identify fraudulent activities and discover crime trends, have also used data mining successfully. For example, these methodologies can aid analysts in the identification of critical behavior patterns, in the communication interactions of narcotics organizations, the monetary transactions of money laundering and insider trading operations, the movements of serial killers, and the targeting of smugglers at border crossings.[5] Data-mining techniques have also been employed by people in the intelligence community who maintain many large data sources as a part of the activities relating to matters of national security. Despite a considerable level of overhype and strategic misuse, data mining has not only persevered but matured and adapted for practical use in the business world.

New Words and Expressions

verify v. 核实，证明，判定
formalize v. 使（协议、计划等）成书面文字形式；使成为正式，使具有一定形式
readily adv. 容易地，乐意地，无困难地
paradigm n. 范例
tremendous adj. 极大的，巨大的，惊人的
data set 数据集
virtually adv. 事实上，几乎，实质上
iterative adj. 迭代的，重复的
interpret v. 解释，翻译

nontrivial adj. 非平凡的
spectrum n. 光谱，频谱，范围
considerably adv. 相当地，非常地
cluster n. 聚集，收集
finite adj. 有限的，限定的
graphical adj. 图表的
elaborate adj. 精心制作的，精巧的，复杂的
collectively adv. 共同地，全体地
intimidating adj. 吓人的
a host of 许多

Notes

[1] In the absence of first-principle models, such readily available data can be used to derive models by estimating useful relationships between a system's variables (i.e., unknown input-output dependencies). 在缺乏基本原理模型的情况下，可用现成的数据通过估计系统变量之间的有用关系（即未知的输入输出依赖关系）来导出模型。

[2] Data mining is most useful in an exploratory analysis scenario in which there are no predetermined notions about what will constitute an "interesting" outcome. 数据挖掘在一个探索性分析场景中最有用，在这个场景中，对什么产生一个"有趣的"结果没有预定的概念。

[3] Current introductory classifications and definitions are given here only to give the reader a feeling of the wide spectrum of problems and tasks that may be solved using data-mining technology. 目前给出的入门级分类和定义，只为读者提供了对数据挖掘技术可解决的问题范围和任务一种认识。

[4] Since data mining is a natural activity to be performed on large data sets, one of the largest target markets is the entire data-warehousing, data-mart, and decision-support community, encompassing professionals from such industries as retail, manufacturing, telecommunications, health care, insurance, and transportation. 由于数据挖掘是一个需要在大数据集上运行的自然活动，其中一个最大的目标市场是整个数据仓库、数据集市和决策支持的社区，同时包括零售、制造、电信、医疗、保险、运输等行业的专业人士。

[5] For example, these methodologies can aid analysts in the identification of critical behavior patterns, in the communication interactions of narcotics organizations, the monetary

transactions of money laundering and insider trading operations, the movements of serial killers, and the targeting of smugglers at border crossings. 例如，这些方法可以帮助分析师进行关键行为模式的识别，如毒品组织的沟通互动，洗钱和内幕交易操作的货币交易，连环杀手的活动，以及追踪边境口岸的走私者。

Unit 10

A Translation of Negative Sentences（否定句的翻译）

英汉两种语言表示否定意义时，在表达方式和用词方面有很大差异。首先，英语的否定形式多种多样，有部分否定、全部否定、几乎否定、双重否定、形式否定等六种，而汉语的否定形式单一。其次，否定句中又按否定部分分类，根据否定部分的不同，又分为一般否定和特殊否定两种。前者是指谓语部分的否定，后者是指除谓语以外部分的否定。所以，一个英语否定句是翻译成否定句还是肯定句，在翻译时需要认真领会原句的含义，才不会发生误译。

1. 形式否定但意义肯定的句子

Example 1：I *cannot but* take care of these building materials.

译文：我必须看管这些建筑材料。

Example 2：I *can't be* thankful enough to you for what you have done for my experiment.

译文：我非常感激你为我的实验所做的一切。

Example 3：Indeed, it is possible that electrons are made of *nothing but* negative electricity.

译文：的确，电子可能就是由负电荷构成的。

Example 4：There is *no* material *but* will deform more or less under the action of force.

译文：在力的作用下，各种材料或多或少都要变形。

Example 5：The importance of the project can *hardly* be exaggerated.

译文：这项工程的重要性怎么夸大都不过分。

Example 6：It is *impossible* to overestimate the value of the invention.

译文：这项发明的价值怎么高估都不过分。

2. 形式肯定但意义否定的句子

Example 1：This equation is *far from* being complicated.

译文：这个方程式一点也不复杂。

Example 2：This transistor radio is so light that you *hardly* notice you are carrying it.

译文：这种晶体管收音机如此轻，几乎感觉不到是带在身边。

Example 3：These elements are shielded so that they are *free from* the influence of magnetic field.

译文：这些元件已加屏蔽，因此不会受磁场影响。

Example 4: *Very few* of everyday things around us are really pure states of matter.

译文：我们周围的日常用品几乎没有是纯态的。

Example 5: The analysis of three-phase circuits is *little more* difficult than that of single-phase circuit.

译文：三相电路的分析比单相电路难不了多少。

Example 6: Good lubrication *keeps* the bearing *from* being damaged.

译文：润滑良好使轴承不受损坏。

3. 部分否定和全部否定

否定句中的 all，every，both 通常是否定的重点。由于否定的不是整体，而是局部，这类否定句也叫部分否定句。对这类句子，若把 not 放在 all 或 every 之前，其部分否定的意义更为明显。

Example 1: All the chemical energy of the fuel is not converted into heat.

或者写成 Not all the chemical of the fuel is converted into heat.

译文：燃料的化学能并不能全部都变成热能。

若要表示完全否定时，则一般写成：

No chemical energy of the fuel is converted into heat.

译文：燃料的化学能都没有转变为热能。

Example 2: All the scientific experiments can not succeed.

译文：这些科学实验未必都能成功。

Example 3: Both of the computer books are not helped.

译文：这两本计算机书不是都有益的。

Example 4: A transformer provides no power of its own.

译文：变压器本身不产生电力。

Example 5: Every color is not reflected back.

译文：并非每种颜色都能反射回来。

4. 双重否定

双重否定一般翻译成肯定句。

Example 1: There is no substance not being made of atoms.

译文：凡物质都是由原子构成的。

Example 2: For practical reasons, it is impossible to maintain this figure in manufacturing without great cost.

译文：由于实际原因，要在制造中保持这一数字就必然要花很大成本。

Example 3: But for air and water, nothing could live.

译文：要不是有空气和水，什么也活不了。

Example 4: Gases cannot be quickly compressed without generating heat.

译文：气体迅速压缩就一定会产生热量。

Example 5: A radar screen is not unlike a television screen.

译文：雷达荧光屏和电视荧光屏没有什么不一样。

5. 否定转移

一般来说，英语否定句的否定范围（scope of negation）从否定词开始，至句子停顿为止。

Example 1：The rocket does not depend on air for its flight.

译文：火箭不依靠空气飞行。

Example 2：You must take care not to damage the machinery.

译文：小心别把机械设备损坏了。

不少英语否定句中的否定词出现的部位，与其否定的对象并不一致，这是常说的否定部位的转移。因此，正确翻译否定句的关键是弄清句子的否定重点（key points of negation），下面是几种否定转移的情况。

（1）当否定句中出现表示程度、方式、地点、时间、原因、目的或频度的状语或状语从句时，尽管形式上是否定谓语的一般否定，但其否定的部位往往已经转移到了这些状语或状语从句上了。例如：从谓语否定转移到方式状语否定，尤其是当句中含有 as、like 等引出方式状语时，not 的否定对象一般都是这些方式状语。请看下面两句：

Example 3：We don't control the circuit with that kind of switch.（not with...）

译文：我们不用那种开关控制电路。

Example 4：These free electrons usually do not move in a regular way.（not in a...）

译文：这些自由电子通常以不规则的方式运动。

有时状语位置的不同也会引起否定含义的变化。

Example 5：The engineers do not agree frequently.

译文：这些工程师的意见不是经常一致的。

但是，上述状语若位于句首，则不在否定范围内，这时否定重点便是句中的谓语。

Example 6：Frequently the engineers do not agree.

译文：这些工程师常常意见不一致。

但是，表示原因、目的的状语常可能有两种理解，即可理解为在否定范围内，也可理解为在否定范围之外。这时要根据上下文来确定否定重点。

Example 7：The engine did not stop because it burnt out.

译文：这台发动机并不是因为烧坏而停止运转的。（不能理解为：这台发动机因为烧坏而没有停止运转。）

Example 8：The computer is not valuable because it is expensive.

译文：计算机不是因为价格昂贵才有价值。（不能理解为：计算机因为价格昂贵才没有价值。）

Example 9：We haven't called the meeting to discuss this question.

译文：我们开会并不是为了讨论这个问题。（也可以理解为：我们没有开会讨论这个问题。）

Example 10：The engineers did not adopt the plan for that reason.

译文：工程师们不是由于那个原因采纳这个方案的。（也可理解为：由于那个原因，工程师们没有采纳这个方案。）

由于上述例句模棱两可，容易误解，因此，写作时要尽量避免。也可以采用下述方法说明此类状语在否定范围之外。

1）在状语前加上表示停顿的逗号。

Example 11：Smoking is not allowed near a store of petrol, because of the danger of an explosion. 译文：因有爆炸危险，不允许在油库附近抽烟。

2）将状语提到句首。

Example 12：For that reason the engineers did not adopt the plan.

译文：由于那个原因，工程师们没有采纳这个方案。

（2）否定主句的谓语转移为否定从句的谓语

Example 1：They don't think that it is necessary to recharge these cells.

译文：他们认为不必将这些电池重新充电。

Example 2：We do not consider conventional PID control is outdated.

译文：我们认为传统的 PID 控制没有过时。

（3）否定主语转移为否定谓语

Example 1：No sound was heard.

译文：没有听到声音。

Example 2：There can be no doubt about the difference between these compounds.

译文：对这些化合物之间的差别不应有任何怀疑。

（4）否定宾语（含介词宾语）转移为否定谓语

Example 1：We know of no effective way to store solar energy.

译文：我们知道没有储藏太阳能的有效方法。

Example 2：The citizen can violate the law on no conditions.

译文：公民任何情况下都不能违反法律。

Exercises

Put the following sentences into Chinese by the translation techniques of negative sentences.

（1）They went to farther than a foreign country.

（2）Supersonic aircraft will not make all subsonic aircraft obsolete.

（3）The previous chapter tells us nothing about what electricity is.

（4）We do not consider melting or boiling to be chemical changes.

（5）Seeing a ball flying, we don't expect the ball to fly forever.

（6）Because of the establishment of the theory of relativity, scientists never again regarded the world as they had before.

（7）After an insulator has been electrified, the electricity does not move through the insulator as it would through a conductor.

（8）The motion is made into a picture by the television camera which does not take a complete picture like any ordinary camera.

（9）In a thermal power plant, all the chemical energy of the coal is not converted into electric power.

(10) Everything is not straightened out.

(11) The experiment on the transformation of energy shows that no energy can be created and none destroyed.

(12) We shall consent to the designing plan under no circumstances.

B   Control of Wind Energy Conversion Systems (风能转换系统控制)

Control plays a very important role in modern wind energy conversion systems (WECS). In fact, wind turbine control enables a better use of the turbine capacity as well as the alleviation of aerodynamic and mechanical loads that reduce the useful life of the installation.[1] Furthermore, with individual large-scale wind facilities approaching the output rating of conventional power plants, control of the power quality is required to reduce the adverse effects on their integration into the network.[2] Thus, active control has an immediate impact on the cost of wind energy. Moreover, high performance and reliable controllers are essential to enhance the competitiveness of wind technology. WECS have to cope with the intermittent and seasonal variability of the wind. By this reason, they include some mechanism to limit the captured power in high wind speeds to prevent from overloading.[3] One of the methods of power limitation basically reduce the blades lift as the captured power approximates its rated value. To this end, the turbines incorporate either electromechanical or hydraulic devices to rotate the blades—or part of them—with respect to their longitudinal axes.[4] These methods are referred to as pitch control ones. Alternatively, there are passive control methods that remove the need for vulnerable active devices, thus gaining in hardware robustness. These methods are based on particular designs of the blades that induce stall at higher than rated wind speed. That is, a turbulent flow deliberately arises at high wind speeds such that aerodynamic torque decreases due to stronger drag forces and some loss of lift.[5] Despite their hardware simplicity, passive stall controlled WECS undergo reduced energy capture and higher stresses that potentially increase the danger of fatigue damage.[6] WECS schemes with the electric generator directly connected to grid have predominated for a long time. In these WECS, the rotational speed is imposed by the grid frequency.[7] Although reliable and low-cost, these fixed-speed configurations are too rigid to adapt to wind variations. In fact, since maximum power capture is achieved at the so-called optimum tip-speed-ratio, fixed-speed WECS operate with optimum conversion efficiency only at a single wind speed. In order to make a better use of the turbine, variable-speed WECS were subsequently developed. They incorporate electronic converters an interface between the generator and alternating current (AC) grid, thereby decoupling the rotational speed from the grid frequency. These WECS also include speed control to track the optimum tip-speed-ratio up to rated speed. Additionally, the electronic converters can be controlled to perform as reactive power suppliers or consumers according to the power system requirements. Fixed-speed pitch-controlled schemes prevailed in early medium to high power wind turbines. Later on, WECS comprising induction generators directly connected to grid and stall-regulated wind rotors dominated the market for many years. More recently, the increasing turbine size and the greater penetration of wind energy into the utility together with exigent standards of power quality were demanding the use of active-controlled configurations. On the one hand, variable-speed schemes finally succeeded, not only because of their increased energy capture but mainly due to their flexibility to improve power quality and to alleviate the loading on the drive-train and tower. On the other hand, the interest in pitch-controlled turbines has lately been reviving due to the tendency towards larger wind turbines, being mechanical stresses an increasing concern as turbines grow in size. By these reasons, variable-speed pitch-controlled wind turbines are currently the preferred option, particularly in medium to high power. In fact, the benefits of control flexibility (e.g., improved power quality, higher conversion efficiency, and longer useful life) largely outweigh the higher complexity and extra initial investments of variable-speed variable-pitch turbines.

In classical gain scheduling techniques, the nonlinear or time-varying plant is linearized around a selected set of operating points and a linear controller is subsequently designed for each of these linear time-invariant (LTI) plants. Then, the gain-scheduled controller is obtained from the family of linear controllers by means of a switching or interpolation algorithm. Gain scheduling techniques have been extensively used by practicing engineers and can be found in a wide range of applications. However, in the absence of theoretical foundations, these techniques come without guarantees. More precisely, stability, robustness and performance properties of the gain-scheduled controlled system cannot be assessed from the feedback properties of the family of LTI control systems. In the early 1990s, Shamma and Athans introduced the linear parameter varying (LPV) systems. LPV models are generally obtained by reformulating a nonlinear or time-varying system as a linear system whose dynamics depend on a vector of time-varying exogenous parameters. In addition to providing a formal framework, the concepts of LPV systems simplify the synthesis of gain-scheduled controllers. In this context, the design task can be formulated as a convex optimization problem with linear matrix inequalities (LMIs). This optimization approach is very effective to solve a wide range of control problems thanks to the existence of efficient numerical algorithms. In LPV gain scheduling techniques based on LMIs optimization the controller is treated as a unique entity, thereby simplifying the scheduling algorithm. In many aspects, the controller design follows a procedure similar to $H\infty$ control, with the difference that the resultant controller is now dependent on the scheduling parameters.[8]

New Words and Expressions

longitudinal   *adj.*   纵的
decouple   *v.*   退偶
intermittent   *adj.*   间歇的
exigent   *adj.*   紧急的

exogenous   *adj.*   外生的
interpolation   *n.*   插值法
vulnerable   *adj.*   易受伤的，脆弱的

Notes

[1] In fact, wind turbine control enables a better use of the turbine capacity as well as the alleviation of aerodynamic and mechanical loads that reduce the useful life of the installation.
事实上，风力涡轮控制不仅能够更好地利用涡轮容量，也能够降低空气动力和机械负

载，而这种负载会缩短装置的使用寿命。

［2］ Furthermore, with individual large-scale wind facilities approaching the output rating of conventional power plants, control of the power quality is required to reduce the adverse effects on their integration into the network. 此外，随着个别大型风力发电设备的输出功率接近传统发电厂，需要电能质量控制来减少联入电网的不利影响。

［3］ WECS have to cope with the intermittent and seasonal variability of the wind. By this reason, they include some mechanism to limit the captured power in high wind speeds to prevent from overloading. 风力发电系统需要处理间歇性和季节性的风力变化，因此，它包括了一些机械装置来限制高风速下获得的电能以防止过载。

［4］ To this end, the turbines incorporate either electromechanical or hydraulic devices to rotate the blades—or part of them—with respect to their longitudinal axes. 为了这个目的，涡轮包含了电子机械或者液压装置使叶片或部分叶片绕纵轴旋转。

［5］ That is, a turbulent flow deliberately arises at high wind speeds such that aerodynamic torque decreases due to stronger drag forces and some loss of lift. 也就是说，在高风速时蓄意产生湍流，以使空气动力转矩因更强的牵引力和部分升力的损失而减小。

［6］ Despite their hardware simplicity, passive stall controlled WECS undergo reduced energy capture and higher stresses that potentially increase the danger of fatigue damage. 尽管硬件简单，被动失速控制的风力发电系统能量获得量低，受到的压力大，从而可能增加设备劳损的风险。

［7］ WECS schemes with the electric generator directly connected to grid have predominated for a long time. In these WECS, the rotational speed is imposed by the grid frequency. 整合了直接连接电网的发电机的风力发电系统在很长一段时间内占据主导地位，在这些风力发电系统中，转速由电网频率决定。

［8］ In many aspects, the controller design follows a procedure similar to $H\infty$ control, with the difference that the resultant controller is now dependent on the scheduling parameters. 在很多方面，控制器设计依据一个和 $H\infty$ 控制相似的步骤，不同在于形成的控制器取决于调度参数。

C Data-Mining Process（数据挖掘过程）

Without trying to cover all possible approaches and all different views about data mining as a discipline, let us start with one possible, sufficiently broad definition of data mining:

Data mining is a process of discovering various models, summaries, and derived values from a given collection of data.

The word "process" is very important here. Even in some professional environments there is a belief that data mining simply consists of picking and applying a computer-based tool to match the presented problem and automatically obtaining a solution.[1] This is a misconception based on an artificial idealization of the world. There are several reasons why this is incorrect. One reason is that data mining is not simply a collection of isolated tools, each completely different from the other and waiting to be matched to the problem. A second reason lies in the notion of matching a problem to a technique. Only very rarely is a research question stated sufficiently precisely that a single and simple application of the method will suffice. In fact, what happens in practice is that data mining becomes an iterative process. One studies the data, examines it using some analytic technique, decides to look at it another way, perhaps modifying it, and then goes back to the beginning and applies another data-analysis tool, reaching either better or different results. This can go around many times; each technique is used to probe slightly different aspects of data—to ask a slightly different question of the data. What is essentially being described here is a voyage of discovery that makes modern data mining exciting. Still, data mining is not a random application of statistical and machine-learning methods and tools. It is not a random walk through the space of analytic techniques but a carefully planned and considered process of deciding what will be most useful, promising, and revealing.

It is important to realize that the problem of discovering or estimating dependencies from data or discovering totally new data is only one part of the general experimental procedure used by scientists, engineers, and others who apply standard steps to draw conclusions from the data.[2] The general experimental procedure adapted to data-mining problems involves the following steps:

1. State the Problem and Formulate the Hypothesis

Most data-based modeling studies are performed in a particular application domain. Hence, domain-specific knowledge and experience are usually necessary in order to come up with a meaningful problem statement. Unfortunately, many application studies tend to focus on the data-mining technique at the expense of a clear problem statement. In this step, a modeler usually specifies a set of variables for the unknown dependency and, if possible, a general form of this dependency as an initial hypothesis. There may be several hypotheses formulated for a single problem at this stage. The first step requires the combined expertise of an application domain and a data-mining model. In practice, it usually means a close interaction between the data-mining expert and the application expert. In successful data-mining applications, this cooperation does not stop in the initial phase; it continues during the entire data-mining process.

2. Collect the Data

This step is concerned with how the data are generated and collected. In general, there are two distinct possibilities. The first is when the data-generation process is under the control of an expert (modeler): this approach is known as a designed experiment. The second possibility is when the expert cannot influence the data-generation process: This is known as the observational approach. An observational setting, namely, random data generation, is assumed in most data-mining applications. Typically, the sampling distribution is completely unknown after data are collected, or it is partially and implicitly given in the data-collection procedure. It is very important, however, to understand how data collection affects its theoretical distribution, since such a priori knowledge can be very useful for modeling and, later, for the final interpretation of results.[3] Also, it is important to make sure that the data used for estimating a model and the data used later for testing and applying a model come from the same unknown sampling distribution. If this is not the case, the estimated model cannot be successfully used in a final application of the results.

3. Preprocess the Data

In the observational setting, data are usually "collected" from the existing data bases, data warehouses, and data marts. Data preprocessing usually includes at least two common tasks:

(1) Outlier Detection (and Moval)

Outliers are unusual data values that are not consistent with most observations. Commonly, outliers result from measurement errors, coding and recording errors, and, sometimes are natural, abnormal values. Such non-representative samples can seriously affect the model produced later. There are two strategies for dealing with outliers:

Detect and eventually remove outliers as a part of the preprocessing phase, or develop robust modeling methods that are insensitive to outliers.

(2) Scaling, Encoding, and Selecting Features

Data preprocessing includes several steps, such as variable scaling and different types of encoding. For example, one feature with the range [0, 1] and the other with the range [-100, 1000] will not have the same weight in the applied technique; they will also influence the final data-mining results differently. Therefore, it is recommended to scale them, and bring both features to the same weight for further analysis. Also, application-specific encoding methods usually achieve dimensionality reduction by providing a smaller number of informative features for subsequent data modeling.

These two classes of preprocessing tasks are only illustrative examples of a large spectrum of preprocessing activities in a data-mining process.

Data-preprocessing steps should not be considered as completely independent from other data-mining phases. In every iteration of the data-mining process, all activities, together, could define new and improved data sets for subsequent iterations. Generally, a good preprocessing method provides an optimal representation for a data-mining technique by incorporating a priori knowledge in the form of application-specific scaling and encoding. [4]

4. Estimate the Model

The selection and implementation of the appropriate data-mining technique is the main task in this phase. This process is not straightforward; usually, in practice, the implementation is based on several models, and selecting the best one is an additional task.

5. Interpret the Model and Draw Conclusions

In most cases, data-mining models should help in decision making. Hence, such models need to be interpretable in order to be useful because humans are not likely to base their decisions on complex "black-box" models. [5] Note that the goals of accuracy of the model and accuracy of its interpretation are somewhat contradictory. Usually, simple models are more interpretable, but they are also less accurate. Modern data-mining methods are expected to yield highly accurate results using high-dimensional models. The problem of interpreting thesemodels (also very important) is considered a separate task, with specific techniques to validate the results. A user does not want hundreds of pages of numerical results. He does not understand them; he cannot summarize, interpret, and use them for successful decision making.

Even though the focus of this chapter is on steps 3 and 4 in the data-mining process, we have to understand that they are just two steps in a more complex process. All phases, separately, and the entire data-mining process, as a whole, are highly iterative, as shown in Fig. 10-1. A good understanding of the whole process is important for any successful application. No matter how powerful the data-mining method used in step 4 is, the resulting model will not be valid if the data are not collected and preprocessed correctly, or if the problem formulation is not meaningful.[6]

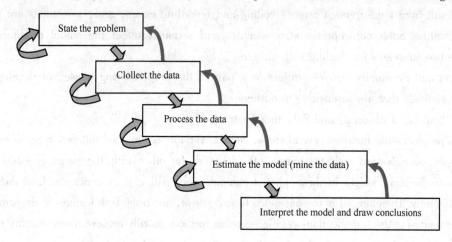

Fig. 10-1　The data-mining process

New Words and Expressions

discipline　n.　学科，纪律	observational　adj.　观测的，根据观察的
derived value　n.　导出值	sampling distribution　n.　抽样分布
precisely　adv.　精确地，恰恰	scale　v.　缩放
go around　供应，(消息流传)	illustrative　adj.　说明的，做例证的
avoyage of discovery　探索未知世界的航行	priori　adj.　先验的
hypothesis　n.　假设	implementation　n.　实现
expense　n.　花费，代价	draw conclusions　得出结论
modeler　n.　模型制造者	contradictory　adj.　有争议的
expertise　n.　专门的知识，专家的意见	high-dimensional　adj.　高维的
distinct　adj.　明显的，清楚的	numerical　adj.　数值的，数字的

Notes

[1] Even in some professional environments there is a belief that data mining simply consists of picking and applying a computer-based tool to match the presented problem and automatically obtaining a solution. 即使在某些专业领域中也有一种信念，即数据挖掘就是通过选择和应用一种基于计算机的工具来匹配所提出的问题并自动获得解决方案。

[2] It is important to realize that the problem of discovering or estimating dependencies from data or discovering totally new data is only one part of the general experimental procedure used by scientists, engineers, and others who apply standard steps to draw conclusions from the data.

PART I　Techniques of EST E-C Translation（科技英语英译汉翻译技巧）

重要的是要认识到数据依赖的发现或估计、全新数据的挖掘只是科学家、工程师以及用标准步骤从数据中得出结论的其他研究者使用的一般实验过程的一部分。

[3] It is very important, however, to understand how data collection affects its theoretical distribution, since such a priori knowledge can be very useful for modeling and, later, for the final interpretation of results. 然而，重要的是理解数据收集如何影响其理论分布，这是由于这样的先验知识对建模以及后续的结果最终解释非常有用。

[4] Generally, a good preprocessing method provides an optimal representation for a data-mining technique by incorporating a priori knowledge in the form of application-specific scaling and encoding. 一般来说，一个好的预处理方法是通过将先验知识融入具体应用的规范描述和编码中，从而提供数据挖掘技术最佳的表示。

[5] Hence, such models need to be interpretable in order to be useful because humans are not likely to base their decisions on complex "black-box" models. 因此，需要解释这些模型以便使用，因为人类不可能把他们的决定建立在复杂的"黑箱"模型上。

[6] No matter how powerful the data-mining method used in step 4 is, the resulting model will not be valid if the data are not collected and preprocessed correctly, or if the problem formulation is not meaningful. 无论步骤4中使用的数据挖掘方法多么强大，如果数据收集和预处理不正确，或者问题公式描述没有意义，产生的模型都不会是有效的。

Unit 11

A　Translation of Complex Sentences（复杂句的翻译）

科技英语中复杂句较多，并多以长句为主，为的是能够更加具体和客观地阐述科学概念或事实。所谓长句是指含有数量较多或相当长的词组定语、词组状语等的简单句，或含有数量较多、相当长的各种从句的主从复合句及等立复合句。一般来说，造成长句的原因有三方面：① 修饰语过多；② 并列成分多；③ 语言结构层次多。翻译长句时，必须抓住全句中心内容，搞清楚各个部分之间的逻辑关系，分清上下层次以及前后关系，然后将英语原文按照汉语特点和表达方式，正确地翻译出原文的意思，不拘泥于原文的形式。而译文的成功与否很大程度上还要取决于译者英、汉两种语言的水平以及对翻译技巧的灵活应用。长句的常用译法一般有以下四种：

1. 顺译法

英语长句的叙述顺序或逻辑关系与汉语相同时，可以按照原文的顺序依次译出。

Example 1：No such limitation is placed on an AC motor; here the only requirement is relative motion, and since a stationary armature and a rotating field system have numerous advantages, this arrangement is standard practice for all synchronous motor rated above a few kilovolt-amperes.

译文：交流电动机不受这种限制，唯一的要求是相对运动，而且由于固定电枢及旋转磁场系统具有很多优点，所以这种安排是所有容量在几千伏安以上的同步电机的标准作法。

Example 2: If parents were prepared for this adolescent reaction, and realized that it was a sign that the child was growing up and developing valuable powers of observation and independent judgment, they would not be so hurt, and therefore would not drive the child into opposition by resenting and resisting it.

译文：如果做父母的对这种青少年的反应有所准备，而且认为这是一个显示出孩子正在成长，正在增强宝贵的观察力和独立的判断力的标志，他们就不会感到如此伤心，所以也就不会因为对此有愤恨和反对的情绪而把孩子推到对立面去。

Example 3: It quite obvious that the organization of such a computer becomes rather complex, since one must insure that a mix-up does not occur among all the various problems and that the correct information is available for the computer when it is called for, but if not, the computation will wait until it is available.

译文：十分明显，这样一台计算机的结构变得相当复杂，因为必须保证它在处理各种各样的问题时都不产生混淆，而且计算机所需的正确信息可随要随有，如果一时没有了，就要等到有了再进行计算。

2. 逆译法

有些英语长句的表达顺序、句子结构与汉语表达习惯不同，甚至完全相反，这时就必须从后面译起，自下而上，逆着英语原文的顺序翻译。

Example 1: The resistance of any length of a conducting wire is easily measured by finding the potential difference in volts between its ends when a known current is following.

译文：已知导线流过的电流，只要测出导线两端电位差的伏特值，就能很容易得出任何长度导线的电阻值。

Example 2: A student of mathematics must become familiar with all the signs and symbols commonly used in mathematics and bear them in mind firmly, and be well versed in the definitions, formulas as well as the technical terms in the field of mathematics, in order that he may be able to build up the foundation of the mathematical subject and master it well for pursuing advanced study.

译文：为了打好数学基础，掌握好数学，以利于深造，一个学数学的学生应该熟悉和牢记数学中常用的记号和符号，精通数学方面的定义、公式和术语。

Example 3: The poor are the first to experience technological progress as a curse which destroys the old muscle-power jobs that previous generations used as a means to fight their way out of poverty.

译文：对于以往几代人来说，旧式的体力劳动是一种用以摆脱贫困的手段，而技术的进步则摧毁了穷人赖以为生的体力劳动，因此首先体验到技术进步之害的是穷人。

Example 4: Insects would make it impossible for us to live in the world; they would devour all our crops and kill our flocks and herds, if it were not for the protection we get from insect-eating animals.

译文：假如没有那些以昆虫为食的动物保护我们，昆虫将吞噬我们所有的庄稼，害死我们的牛羊家畜，使我们不能生存于世。

3. 分译法

有时英语中的主要成分和短语等所叙述的内容，关系并不密切，各具有相对独立的意义。这时就应该按照汉语常用短句的习惯，把长句中的从句或短语分别译成独立的句子，顺序基本保持不变。有时为了使语气连贯，需加适当的词语。

Example 1：This kind of two-electrodes tube consists of a tungsten filament, which gives off electrons when it is heated, and a plate toward which the electrons migrate when the field is in the right direction.

译文：这种二极管由一根钨丝和一个极板组成，钨丝受热时释放电子，当电场方向为正时，电子就移向极板。

Example 2：Television, it is often said, keeps one informed about current events, allows one to follow the latest developments in science and politics, and offers an endless series of programmes which are both instructive and entertaining.

译文：人们常说，通过电视可以了解时事，掌握科学和政治的最新动态。从电视里还可以看到层出不穷、既有教育意义又有娱乐性的新节目。

Example 3：On account of the accuracy and ease with which resistance measurements may be made and the well-known manner in which resistance varies with temperature, it is common to use this variation to indicate changes in temperature.

译文：我们知道，电阻是随温度而变化的。电阻的测量既准确，又方便，因此通常都用电阻的变化来表示温度的变化。

4. 合译法

在翻译科技英语长句时，对于每一个英语句子的翻译，有时并不只是使用一种翻译方法，而是多种翻译方法的综合运用，这在英语长句的翻译中表现得尤为突出。比如下面例句就是用了顺译法和分译法。

Example 1：The computer performs a supervisory function in the liquid-level control system by analyzing the process conditions against desired performance criteria and determining the changes in process variables to achieve optimum operation.

译文：在液位控制系统中，计算机执行着一种监控功能。它根据给定的特性指标来分析各种过程条件，并决定各过程变量的变化获得最佳操纵。

再分析一下采用了逆译法和分译法的例句。

Example 2：Rocket research has confirmed a strange fact which had already been suspected there is a "high temperature belt" in the atmosphere, with its center roughly thirty miles above the ground.

译文：人们早就怀疑，大气层中有一个"高温带"，其中心在距离地面约30mile（1mile≈1609.34m）的高空。利用火箭进行研究后，这一奇异的事实已得到证实。

Exercises

Put the following sentences into Chinese by the translation techniques of complex sentences.

（1）Since the molecules of a gas are much too far from each other to repel or attract each other, it is very easy compass a gas, while a solid or liquid is almost incompressible, because the repulsions of the electrical charges of which its atoms are made up are far stronger than any force we can apply.

(2) Aluminum remained unknown until the nineteenth century, because nowhere in nature is it found free, owing to its always being combined with other elements, most commonly with oxygen, for which it has a strong affinity.

(3) The method normally employed for free electrons to be produced in electron tubes is thermionic emission, in which advantage is taken of the fact that, if a solid body is heated sufficiently, some of the electrons that it contains will escape from its surface into the surrounding space.

(4) We define an inertial reference system as one relative to which a body does remain at rest or move uniformly in a straight line when no force acts on it.

(5) Up to the present time, throughout the eighteenth and nineteenth centuries, this new tendency placed the home in the immediate suburbs, but concentrated manufacturing activity, business relations, government, and pleasure in the centers of the cities.

(6) Modern scientific and technical books, especially textbooks, require revision at short intervals if their authors wish to keep pace with new ideas, observations and discoveries.

(7) Today they are using supercomputers to build virtual realities that let users explore artificial words existing only in the computer's electronic circuitry.

(8) An application server can further improve on the client-server architecture by moving much or all of the application logic from the client into the middle tier, so that clients concentrate exclusively on presentation logic.

(9) Instead of becoming a specialist in a certain subject and working in that area for a lifetime, you will have to adapt and possibly retrain yourself several times.

(10) The reader might speak into a telephone where the information is transduced from patterns of compressed air molecules traveling at the speed of sound into electronic pulses traveling down a copper wire closer to the speed of light.

B Thermal Power Plant Simulation and Control（热电站仿真与控制）

Almost all coal, nuclear, geothermal, solar thermal electric, and waste incineration plants, as well as many natural gas power plants are thermal. Natural gas is frequently combusted in gas turbines as well as boilers. The waste heat from a gas turbine can be used to raise steam, in a combined cycle plant that improves overall efficiency.[1] Power plants burning coal, oil, or natural gas are often referred to collectively as fossil-fuel power plants. Some biomass-fueled thermal power plants have appeared also. Non-nuclear thermal power plants, particularly fossil-fueled plants, which do not use co-generation are sometimes referred to as conventional power plants.

Commercialelectric utility power stations are most usually constructed on a very large scale and designed for continuous operation. Electric power plants typically use three-phase or individual-phase electrical generators to produce alternating current (AC) electric power at a frequency of 50 Hz or

PART I Techniques of EST E-C Translation （科技英语英译汉翻译技巧）

60 Hz (hertz, which is an AC sine wave per second) depending on its location in the world. [2] Other large companies or institutions may have their own usually smaller power plants to supply heating or electricity to their facilities, especially if heat or steam is created anyway for other purposes. Shipboard steam-driven power plants have been used in various large ships in the past, but these days are used most often in large naval ships. Such shipboard power plants are general lower power capacity than full-size electric company plants, but otherwise have many similarities except that typically the main steam turbines mechanically turn the propulsion propellers, either through reduction gears or directly by the same shaft. The steam power plants in such ships also provide steam to separate smaller turbines driving electric generators to supply electricity in the ship. Shipboard steam power plants can be either conventional or nuclear; shipboard nuclear plants are with very few exceptions only in naval vessels. There have been perhaps about a dozen turbo-electric ships in which a steam-driven turbine drives an electric generator which powers an electric motor for propulsion.

In some industrial, large institutional facilities, or other populated areas, there arecombined heat and power (CH&P) plants, often called co-generation plants, which produce both power and heat for facility or district heating or industrial applications. AC electrical power can be stepped up to very high voltages for long distance transmission with minimal loss of power. Steam and hot water lose energy when piped over substantial distance, so carrying heat energy by steam or hot water is often only worthwhile within a local area or facility, such as steam distribution for a ship or industrial facility or hot water distribution in a local municipality.

The energy efficiency of a conventional thermal power station, considered as salable energy (in MW) produced at the plant busbars as a percent of the heating value of the fuel consumed, is typically 33% to 48% efficient. This efficiency is limited as all heat engines are governed by the laws ofthermodynamics. The rest of the energy must leave the plant in the form of heat. This waste heat can go through a condenser and be disposed of with cooling water or in cooling towers. If the waste heat is instead utilized for district heating, it is called co-generation. An important class of thermal power station are associated with desalination facilities; these are typically found in desert countries with large supplies of natural gas and in these plants, freshwater production and electricity are equally important co-products.

Since the efficiency of the plant isfundamentally limited by the ratio of the absolute temperatures of the steam at turbine input and output, efficiency improvements require use of higher temperature, and therefore higher pressure, steam. Historically, other working fluids such as mercury have been experimentally used in a mercury vapor turbine power plant, since these can attain higher temperatures than water at lower working pressures. However, the obvious hazards of toxicity, and poor heat transfer properties, have ruled out mercury as a working fluid.

Control system that is central in determining the overall behaviour of the generating unit. All the main control loops must respond to a central command structure, which sets their individual setpoints and controls the behaviour of the plant. [3] It is the demand for steam that resides at the top of this control hierarchy. From this all other individual loop controllers receive their demand or setpoint signal. Due to its importance, the steam demand signal is often known as the master control signal.

The strategic behaviour of the unit is governed by various boiler control configurations, and the behaviour of the master control signal within these arrangements is now discussed.

1) *Boiler following mode.* Boiler following or "constant pressure" mode utilizes the main steam governor as a fast-acting load controller, since opening the governor valves, and releasing the stored energy in the boiler, meets short-term increases in electrical demand.[4] Conversely, closing the governor valves reduces the generated output. These actions alter the main steam pressure, so it is the role of the master pressure controller to suitably adjust the fuel-firing rate.[5] Operating a unit in this mode does, however, contain inefficiencies as throttling of the governor valves reduces the available steam flow, creating energy losses.

2) *Turbine following mode.* A generating unit may alternatively be configured to operate in turbine following mode, whereby the combustion controls of the boiler are set to achieve a fixed output.[6] The position of the main steam governor valve is controlled by the valve outlet pressure, not the input as in boiler following.[7] Consequently, such units can be operated with their governor valves remaining fully open. Turbine following mode is preferred for thermal base load and nuclear plant, since it allows the generating units to operate continuously at their maximum capacity rating. However, such units do not respond to frequency deviations and so cannot assist in a network frequency support role. For nuclear plant, there are also safety benefits in providing continuous steady state operating conditions.

3) *Sliding pressure mode.* Although boiler following mode is commonly used, sliding pressure mode is an "instructive" development, where the constant steam pressure is replaced by a variable steam pressure mode. The reduced throttling back action by the governor control valves, at lower outputs, leads to improved unit efficiency. Variable pressure operation also provides faster unit loading, and enables operation of the turbines at lower temperatures and pressures. However, the ability to use the stored energy of the boiler to meet short-term changes in demand is restricted. For safety reasons, fast-responding, electrically operated safety valves are essential for variable pressure operation to protect against sudden, dangerous increases in steam pressure that may occur while the pressure setpoint is low.

New Words and Expressions

incineration	n.	焚化	throttling	n.	节流
desalination	n.	海水淡化	shipboard	n.	船载
combust	n.	燃烧	thermodynamics	n.	热力学
mercury	n.	汞	municipality	n.	自治市
collectively	n.	集体地			

Notes

[1] The waste heat from a gas turbine can be used to raise steam, in acombined cycle plant that improves overall efficiency. 在联合循环发电厂中，燃气轮机的废气能用来提升蒸汽从而提高总体效率。

[2] Electric power plants typically usethree-phase or individual-phase electrical generators to produce alternating current (AC) electric power at a frequency of 50 Hz or 60 Hz depending on its location in the world. 发电厂通常使用三相或者单相发电机产生频率50Hz或者60Hz的交流电，这取决于它在地球上的位置。

[3] All the main control loops must respond to a central command structure, which sets their individual setpoints and controls the behaviour of the plant. 所有的主要控制回路必须对中央控制命令机构做出响应，这个机构确定它们各自的设定值且控制发电厂的运行特性。

[4] Boiler following or "constant pressure" mode utilizes the main steam governor as a fast-acting load controller, since opening the governor valves, and releasing the stored energy in the boiler, meets short-term increases in electrical demand. 因为打开调节阀，放出锅炉中储存的能量，能够满足电力需求的短期增长，所以锅炉跟随模式或"恒压"模式利用主蒸汽调节器作为一个快速响应负载控制器。

[5] These actions alter the main steam pressure, so it is the role of the master pressure controller to suitably adjust the fuel-firing rate. 这些动作改变了主蒸汽压，所以主压力控制器的作用是适当地调整燃料消耗速率。

[6] A generating unit may alternatively be configured to operate in turbine following mode, whereby the combustion controls of the boiler are set to achieve a fixed output. 可能另外配置以涡轮机跟随模式运行的发电单元，由此设立锅炉的燃烧控制以获得一个固定的输出。

[7] The position of the main steam governor valve is controlled by the valve outlet pressure, not the input as in boiler following. 主蒸汽调节阀的位置由阀门出口压力控制，而不是锅炉跟随模式中的输入。

C Industry 4.0: the Fourth Industrial Revolution（工业4.0：第四次工业革命）

The industrial sector is important to the EU economy and remains a driver of growth and employment. Industry (which in this context means manufacturing and excludes mining, construction and energy) provides added value through the transformation of materials into products. Although only roughly 1 in 10 enterprises in the EU is classified as manufacturing, the sector comprises 2 million companies and is responsible for 33 million jobs. It is also responsible for over 80% of exports and 80% percent of private research and innovation, and as such is one of the key elements of sustainable economic growth. Moreover, every new job in manufacturing results in the creation of between one half and 2 jobs in other sectors.

However the relative contribution of industry to the EU economy is declining. The European economy has lost a third of its industrial base over the past 40 years. By the third quarter of 2014, the value added by manufacturing to the economy in the EU represented only 15.3% of total value added, a decline of 1.2 percentage points since the beginning of 2008. This "de-industrialisation", a process which is also present in other developed economies, is in part due to the rise of manufacturing in other parts of the world, the relocation of labour-intensive work to countries with lower labour costs and global supply chains with suppliers located outside the EU. Moreover, the growing services sector represents an ever-larger proportion of the total European economy, resulting in a lower relative share for industry.

In 2012, in response to this decline in the relative importance of industry, the European Commission set a target that manufacturing should represent 20% of total value added in the EU by 2020. Whilst some observers find this goal overly ambitious, many believe that we are on the brink of a new industrial revolution, Industry 4.0, which could boost the productivity and value added of European industries and stimulate economic growth. As part of its new Digital Single Market Strategy, the European Commission wants to help all industrial sectors exploit new technologies and manage a transition to a smart, Industry 4.0 industrial system.

1. What Is Industry 4.0?

Industry 4.0 is a term applied to a group of rapid transformations in the design, manufacture, operation and service of manufacturing systems and products. The 4.0 designation signifies that this is the world's Fourth Industrial Revolution, the successor to three earlier Industrial Revolutions that caused quantum leaps in productivity and changed the lives of people throughout the world.[1] In the words of German Chancellor Angela Merkel, Industry 4.0 is "the comprehensive transformation of the whole sphere of industrial production through the merging of digital technology and the Internet with conventional industry". In short, everything in and around a manufacturing operation (suppliers, the plant, distributors, even the product itself) is digitally connected, providing a highly integrated value chain. The term "Industry 4.0" originated in Germany, but the concept largely overlaps developments that, in other European countries, may variously be labelled: Smart factories, the Industrial Internet of Things, smart industry, or advanced manufacturing.

Industry 4.0 depends on a number of new and innovative technological developments:

1) The application of information and communication technology (ICT) to digitise information and integrate systems at all stages of product creation and use (including logistics and supply), both inside companies and across company boundaries.[2]

2) Cyber-physical systems that use ICTs to monitor and control physical processes and systems. These may involve embedded sensors, intelligent robots that can configure themselves to suit the immediate product to be created, or additive manufacturing (3D printing) devices.

3) Network communications including wireless and Internet technologies that serve to link machines, work products, systems and people, both within the manufacturing plant, and with suppliers and distributors.

4) Simulation, modelling and virtualisation in the design of products and the establishment of manufacturing processes.

5) Collection of vast quantities of data, and their analysis and exploitation, either immediately on the factory floor, or through big data analysis and cloud computing.

6) Greater ICT-based support for human workers, including robots, augmented reality and intelligent tools.

Industry 4.0 is expected to have a major effect on global economies. Industry 4.0 can deliver estimated annual efficiency gains in manufacturing of between 6% and 8%. The Boston Consulting Group predicts that in Germany alone, Industry 4.0 will contribute 1% per year to GDP over ten years, creating up to 390,000 jobs. Globally, one expert estimates that investment on the Industrial Internet will grow from US $ 20 billion in 2012 to more than US $ 500 billion in 2020 (albeit with slower growth after that date), and that value added will surge from US $ 23 billion in 2012 to US $ 1.3 trillion in 2020.

Unsurprisingly, Europe is not the only region of the world to take an interest in digital manufacturing. The United States has established a National Network for Manufacturing Innovation with a proposed US $ 1 billion of public funding to bring together national research centres investigating topics such as digital manufacturing and design. Companies in the Asia/Pacific region were expected to invest almost US $ 10 billion in the Industrial Internet of Things in 2012, with that figure rising to nearly US $ 60 billion by 2020. If the EU is to remain competitive and to reach its goal of becoming a smart, sustainable and inclusive economy by 2020, European industry will need to capture the al for productivity and growth that Industry 4.0 appears to offer.[3]

2. What will Industry 4.0 Change?

Digitalised manufacturing will result in a wide range of changes to manufacturing processes, outcomes and business models. Smart factories allowincreased flexibility in production. Automation of the production process, the transmission of data about a product as it passes through the manufacturing chain, and the use of configurable robots means that a variety of different products can be produced in the same production facility.[4] This mass customisation will allow the production of small lots (even as small as single unique items) due to the ability to rapidly configure machines to adapt to customer-supplied specifications and additive manufacturing. This flexibility also encourages innovation, since prototypes or new products can be produced quickly without complicated retooling or the setup of new production lines.

Thespeed with which a product can be produced will also improve. Digital designs and the virtual modelling of manufacturing process can reduce the time between the design of a product and its delivery. Data-driven supply chains can speed up the manufacturing process by an estimated 120% in terms of time needed to deliver orders and by 70% in time to get products to market.

Integrating product development with digital and physical production has been associated with large improvements inproduct quality and significantly reduced error rates. Data from sensors can be used to monitor every piece produced rather than using sampling to detect errors, and error-correcting machinery can adjust production processes in real time. This data can also be collected and analysed using "big data" techniques to identify and solve small but ongoing problems. The rise in quality plays an important role in reducing costs and hence increasing competitiveness: The top 100 European manufacturers could save an estimated 160 billion in the costs of scrapping or reworking defective products if they could eliminate all defects.

Productivitycan also increase through various Industry 4.0 effects. By using advanced analytics in predictive maintenance programmes, manufacturing companies can avoid machine failures on the factory floor and cut downtime by an estimated 50% and increase production by 20%. Some companies will be able to set up "lights out" factories where automated robots continue production without light or heat after staff has gone home. Human workers can be used more effectively, for those tasks for which they are really essential. For example, in the Netherlands, Philips produces electric razors in a "dark factory" with 128 robots and just nine workers, who provide quality assurance.

Customerswill be able to be more involved in the design process, even supplying their own modified designs which can then be quickly and cheaply produced. The location of some manufacturing operations may also be close to the customer: If manufacturing is largely automated, it does not need to be "off-shored" or located in distant countries with low labour (but high transport) costs. European companies may decide to bring some manufacturing capacity back to Europe ("re-shore"), or to establish new plants in Europe rather than abroad.

Industry 4.0 will also provoke changes inbusiness models. Rather than exclusively competing on cost, European companies can compete on the basis of innovation (the ability to deliver a new product rapidly), on the ability to produce customer-driven customised designs (through configurable factories), or on quality (the reduction of faults due to automation and control).[5] Some companies may take advantage of the data created as "smart" products are created and used, and adopt business models based on selling services not products (sometimes described as "selling light not light bulbs"). This "servitisation" can help to expand business opportunities and increase revenues.

New Words and Expressions

EU (European Union)　欧洲联盟　　　　　augmented reality　增强现实
ICT (information and communication technology)　　mass customisation　大规模定制化
　　信息与通信技术　　　　　　　　　　retool　v.　更换，重新安排，重组
cyber-physical system　信息物理系统　　　provoke　v.　引发，激起，引起，挑衅

Notes

[1] The 4.0 designation signifies that this is the world's Fourth Industrial Revolution, the successor to three earlier Industrial Revolutions that caused quantum leaps in productivity and changed the lives of people throughout the world. 4.0 这个名称指的是世界第四次工业革命，以前的三次工业革命引起了生产格局的突变，并改变了全世界人类的生活。

[2] The application ofinformation and communication technology (ICT) to digitise information and integrate systems at all stages of product creation and use (including logistics and supply), both inside companies and across company boundaries; 将信息与通信技术应用于数字化信息，并从产品创作和使用（包括物流和供货）的各阶段在公司内部以及跨公司集成系统。

[3] If the EU is to remain competitive and to reach its goal of becoming a smart, sustainable and inclusive economy by 2020, European industry will need to capture the potential for productivity and growth that Industry 4.0 appears to offer. 如果欧盟计划在 2020 年保持竞争力，并达到使其成为一个智慧、可持续并且包容的经济体的目标，欧洲工业就需要抓住工业 4.0 所提供的生产和增长潜在力量。

PART I　Techniques of EST E-C Translation（科技英语英译汉翻译技巧）

[4] Automation of the production process, the transmission of data about a product as it passes through the manufacturing chain, and the use of configurable robots means that a variety of different products can be produced in the same production facility. 生产过程自动化，产品通过生产链的数据传输，组合机器人的使用都意味着各种不同的产品能够在相同的生产设施系统中生产。

[5] Rather than exclusively competing on cost, European companies can compete on the basis of innovation (the ability to deliver a new product rapidly), on the ability to produce customer-driven customised designs (through configurable factories), or on quality (the reduction of faults due to automation and control). 欧洲公司不是专注于成本竞争，而是专注于能够在基础创新（快速发布新产品的能力）、用户驱动的定制设计能力（通过工厂配置）或者质量（由自动化和控制减少故障）这些方面具有竞争力。

Unit 12

A　Translation of Numerals（数的翻译）

科技文献中经常会出现表示数量的词语，翻译错了就会产生严重的后果。汉语和英语的数量词语表达方法有不少差异，因此在翻译时必须谨慎对待，避免差错，以期正确无误。

1. 单位问题

不同地区或者国家采用的数字单位不尽相同，美国、法国以及多数欧洲国家采用千位制（thousand system），每 10^3 增设一个单位，常用分节符（,）分开；英国、德国等采用百万位制（million system），即每 10^6 增设一个新单位，而按汉语的习惯规则是每增大 10^4（即一万）设一个新单位。通常我们看到的科技资料绝大多数是采用千位制的，但遇到较大的单位，如 billion（10^9，十亿）、trillion（10^{12}，兆）、decillion（10^{33}）等时，应充分注意计数单位。若是美制就分别是 10^9（十亿）、10^{12}（万亿、兆）、10^{33}；若为英制，则变为 10^{12}（万亿）、10^{18}（百万亿）等，差别非常大。还必须注意汉语中的"十亿"分别对应于美制的"billion"和英制的"one thousand million"。因此，翻译时需要知道文献出自英国还是美国，再进行换算，否则容易出错。

Example 1：Petroleum, natural gas, and coal come from plants that lived 330 million years ago.

译文：石油、天然气和煤都是由三亿三千万年以前的植物形成的。

Example 2：Statistics already calculate a world population of 8.7 milliard for the year 2050. (milliard 为英语中的十亿，等于美制中的 billion)

译文：统计数字已经算出，2050 年的世界人口数是 87 亿。

2. 数词复数形式的翻译方法

英语中某些数词可以有复数形式，并可与 of 短语连用，用来表示一种不确定的约数。

常见的有 tens（数十）、dozens（几十，许多）、scores of（好几十，大量）、decades（几十年）、hundreds（数百）、thousands（数千，成千上万）、millions（数百万，千百万）等。

Example 1：Although there exist only about 90 elements in nature, they combine to form *thousands of* different substances in the world.

译文：虽然自然界中大约有 90 种元素，但它们结合起来可以构成世界上<u>成千上万</u>种不同的物质。

Example 2：*Thousands and tens of thousands of* machine parts are made up of simple geometric figures such as squares, circles, triangles, and similar shapes.

译文：<u>千千万万</u>的机械零件都是由简单的几何形状组成的，如方形、圆形、三角形以及其他类似的形状。

3. 倍数的译法

英语中表示倍数可用 N times（两倍为 twice/double/duple，三倍可用 3 times/triples/treble；四倍可用 4 times/quadruple），N-fold，by a factor of N 等。其基本意义是：用于增加时为"乘"，用于减少时是"除"。

常见的句型有：

1）N times + the（或物主代词）+ 名词

N times + that + 后置定语（多为"of"短语）

N times + as + 形容词/副词原级 + as + 名词

N times + what 从句

2）表示增减的词 +（by）N times/N-fold/by a factor of N

3）as much（many…）as 或 again as… as

第三种结构中的副词 again 表示"再一倍"的意思。通常译作"是……两倍"，"两倍于……"或"比……大（多）一倍"。

（1）数量增加的译法

汉语在表示倍数增加时要区别是否包括底数在内的问题，即汉语倍数往往取决于前面所用的动词。例如："甲是乙的 N 倍"，"甲增加到乙的 N 倍"，都包括基数，也就是说净增 N-1 倍。而"甲比乙增加了 N 倍"则不包括基数，也就是说净增 N 倍。而英语不管采用何种表达法，倍数都包括基数。在英译汉时，倍数的增加可用两种方法来表示：

1）"增加了 N-1 倍"（不包括基数）。

2）"增加到 N 倍"，或"是原来的 N 倍"（包括基数）。

Example 1：This machine is *five times* heavier than that one.

译文：这台机器比那台重<u>四倍</u>。（或：这台机器是那台机器的<u>五倍重</u>。）

Example 2：In case of electronic scanning the bandwidth is broaden by *a factor of two*.

译文：如果用电子扫描，带宽需要<u>增加一倍（增加到两倍）</u>。

Example 3：By 1978, a second generation of microprocessors had become available that are about 10 *times* as powerful as their predecessors.

译文：到 1978 年出现了第二代微处理器，其计算能力约为第一代的<u>10 倍</u>。

Example 4：Multiband transmission permits a reduction in error probability in exchange for at least *a twofold* increase in bandwidth and carrier power.

译文：宽频带传输能降低误差概率，但其代价是带宽和载波功率至少增加<u>一倍</u>。

Example 5：The machine improves the working conditions and raises efficiency *four times*.

译文：这台机器改变了工作条件，并使之效率提高了<u>三倍</u>。

Example 6：The speed exceed the average speed by *a factor of* 2.5.

译文：该速度超过了平均速度<u>1.5倍</u>。

Example 7：The leads of the new component are *as long as again* as those of the old.

译文：新型元件的引出线长度比旧式的<u>加长了一倍</u>。

但是，如果倍数是一个很大的近似值，相差一倍并无多大影响时，则无须减去1。

Example 8：The thermal conductivity of metals is as much as *several hundred times* that of glass.

译文：金属的导热率比玻璃高<u>数百倍</u>。

（2）数量减少的译法

按照汉语的习惯，用于减少不可用"倍"。英语中的 times 用于减少时的基本意义是"除"，其结果相当于汉语中的分数。因此 N times 用于减少时可译为"减少到 1/N"，或"减少了（N-1）/N"。此外，汉语的分母不习惯用小数，若英语的倍数为小数，则应化为整数进行换算。例如：reduce 3.5 times，应换算成 1/3.5 = 2/7，译成"减少到七分之二"或"减少了七分之五"。

Example 1：The equipment under development till reduce the error probability *by a factor of* 7.

译文：正在研制的设备将使误差减小到 1/7。

Example 2：The switching time of the new type transistor is <u>shortened five times</u>.

译文：这种新型晶体管的开关时间<u>缩短了 4/5</u>。

Example 3：The power output of this small electric machine is <u>three times less than its input</u>.

译文：这台微电机的输出功率<u>仅为输入功率的 1/3</u>。

Example 4：The error probability of the equipment <u>was reduced by 2.5 times</u> through technical innovation.

译文：通过技术革新，该设备的误差概率降低了 3/5。

此外，表示减少的含义还有以下常用的几种：

    decrease one-half　　减小一半

    reducing... by one-half　　减小一半

    cut... in half　　把……减小一半

    shorten... two times　　缩短一半

    twice less　　少了一半

    one-half less　　少了一半

4. 分数和百分数的译法

分数和百分数无论增减，一般都表示净增减的部分，不包括基数在内，可以照译不变。

Example 1：In this lathe, proper lubrication has diminished almost *three fifths* of the friction.

译文：把这台车床适当加油润滑，可使摩擦力几乎减少 3/5。

Example 2：A temperature rise of one degree centigrade raises the electric conductivity of a semiconductor by 3%-6%.

译文：温度升高 1℃，半导体的导电率就增大 3%~6%。

5. 表示数量的习惯短语的译法

英语中有些习惯短语，表示数量，在科技文献中常见。

Example 1：*Ten to one*, forms of energy come from the sun directly or indirectly.

译文：十之八九，能量的形式直接或间接来自太阳。

Example 2：In fact, *a hundred and one* (or, *a thousand and one*) mechanical devices waste a great deal of energy in overcoming friction.

译文：事实上，许多的机械装置在克服摩擦过程中浪费大量的能量。

Example 3：The tungsten carbide is hard material *second to none* among metals.

译文：碳化钨是金属中首屈一指的硬材料。

Example 4：It is expected that the problem can be solved by the electronic computer of *the order of* two minutes.

译文：预计这个问题用电子计算机约两分钟就能解决。

此外，还有几个关于"... + every + numbers + noun +..."结构的译法：

every other = each second（每隔一个，每两个）
every three days = every third day（每隔两个，每三个）
at 10-day intervals = every 10 days = every tenth day（每隔九个，每十个）

Exercises

Put the following sentences into Chinese by the translation techniques of numerals.

(1) Machining is not an economical method of producing a shape, because it has good raw materials converted into hundreds of thousands of scrap chips.

(2) The drain voltage has been increased by a factor of four.

(3) The presence of the iron in the coil has increased the magnetic induction to over 5,500 times what it would be if the coil were vacuum.

(4) The production of integrated circuits has been increased to three times as compared with last year.

(5) This year the output of electric machine has gone up twice over that of last year.

(6) The input is again as great as the output.

(7) The bed of this lathe is half as long again as that of the other one.

(8) The sun is 330,000 times more massive and a million times more bulky than the earth.

(9) The automatic assembly line shortened the assembling time five times.

(10) If you treble the distance, the gravitational attraction gets nine times weaker.

(11) This substance reacts three times as fast as the other one.

(12) The capacitance will decrease by 750 parts per million per each degree rise in temperature.

B Chasing the Clouds—Distributed Computing and Small Business
（追逐云——分布式计算与小型商业）

As a young boy, I always wondered what it would be like to be a weatherman on TV. I'd sit with my mom and dad, watching the local news from San Francisco and marveling at the way Pete Giddings on Channel 7 always seemed to know exactly what was going to happen. I thought to myself that this would be a cool thing—to know the future ahead of time! More than two decades later, I actually did get the opportunity to be a TV weatherman, though just as a fill-in for the vacationing regular on a tiny station on Cape Cod, Massachusetts in my former life as a sports reporter. [1] Of course, by then, I realized that there wasn't so much magic in the forecasting, but a lot of science. I still was fascinated by the science of weather, though, and especially found the variety and function of clouds to be most interesting.

Now, about 25 years later, clouds have reached a whole new level of significance in my life as a technology advisor and consultant. Like the effect of the naturally-occurring namesakes in the sky, the emergence of the cloud in the computing world seems to have signaled a change in the climate of business computing. But just what is this notion? And what are the good and bad consequences of adopting this growing trend to your business, especially with the tenuous economy and the more pronounced effects on the small to midsized business. [2]

First of all, we need to understand exactly what is "cloud computing". Since the early days of ENIAC (electronic numerical integrator and calculator), man has thirsted for ways to get more and more processing power to perform increasingly complex tasks. Early machines relied on faster internal workings to do this, until the trend became combining many different machines in a cluster on one site to maximize the load sharing. [3] But in the 1990's, Ian Foster and Carl Kesselman had a different take. They postulated that setting up a "grid" over many machines connected remotely to each other could be a better way to share the load over many more CPUs. Unfortunately, since the machines could be separated by huge distances, delays in accessing the data stores or latency kept the concept from gaining any major traction in the business world.

So the acceleration of the Internet and its adoption allowed for a logical next step into "cloud" computing. It is thought that the phrase comes from the long-used symbol in flow charts and diagrams that used a cloud to designate the Internet. Thus, cloud computing takes the grid computing concept and adds a level of decentralization inherent in what the Internet is. Cloud computing can be both private and public. Private clouds are built, used and maintained by an individual organization for internal use or private resale. Public cloud computing is sells services to anyone on the Internet. At this time, the Amazon Web Services called S3 (Simple Storage Service) is the largest public cloud provider, though Microsoft's Azure platform is gaining rapidly, as well as EMC.

The public cloud marketplace has been divided into three types by some cloud model—Software-as-a-Service (SaaS), Infrastructure-as-a-Service (IaaS), and Platform-as-a-Service (PaaS). In the SaaS cloud model, the vendor supplies the hardware infrastructure, the software product and interacts with the user through a front-end portal. SaaS is a very broad market. Services can be anything from Web-based e-mail to inventory control and database processing. IaaS, including Amazon's S3, provides virtual server instances with unique IP addresses and blocks of storage on demand. Customers use the provider's application program interface (API) to start, stop, access and configure their virtual servers and storage. The third menu item, PaaS, is a set of software and product development tools hosted on the provider's infrastructure. This allows developers to create applications on the provider's platform over the Internet and not build and maintain costly testing environments. PaaS providers may use APIs, website portals or gateway software installed on the customer's computer. Force. com, developed by Salesforce. com, and GoogleApps are examples of PaaS. Potential developer customers should be aware that currently, there are no standards for interoperability or data portability in the cloud. Some of the providers will not allow software created by their customers to be moved off the provider's platform. Kind of defeats the purpose, doesn't it?

Recently, enhancements to several key factors have accelerated the shared computing concept to what cloud computing is today. The ever-growing availability of higher-speed connectivity, along with advances in data storage and increased input/output efficiency has allowed this previous notion of grid computing to evolve.[4] Today, cloud-based applications and storage are being touted as the next wave of business tools to leapfrog us well past where we are today. Many larger enterprises have already leveraged the concepts of the cloud to realize great savings, as describe in the applications in the previous paragraph.

How does the cloud create these savings and improvements? There are several technologies and applications that come into play here. One is the ever-growing adoption of virtualization in the data center. Taking many, or even thousands, of servers out of service, replaced by "virtual" servers in a larger host environment is becoming more and more the standard by which data centers are being built or rebuilt.[5] Do the math—fewer servers means less equipment and power to run it, and smaller rooms requiring less cooling. All of that costs money, so there are built-in savings. In addition, with less hardware come fewer points of failure and lower licensing purchases and renewals. Not to mention enhanced disaster recovery with more robust offsite storage options.

While setting up an internal cloud structure is a preferred methodology for some, there are organizations large and small that have opted for an even more radical shift to the third party "cloud services" providers, like Amazon, etc. In many cases, why go through the expense of building your own if you can get all the benefits for a monthly fee? There are of course trade-offs in that using the Internet does expose the data and applications to the potential of breaches, but more secure tunneling methods are making those worries minimal. And with more powerful endpoints, computing power at the desktop can handle the load.

Of course, it makes sense for the largest enterprises and government entities to be embracing these technologies in the tighter economic situation we live in. But the real question is, can the SMB marketplace ride their coattails? The answer is "definitely, maybe". Sure, also the title of a completely mediocre recent romantic comedy in the theaters, but also indicative about the split between expert opinions. A Forrester Research study in May of 2009 reported that an estimated 2% of small businesses (categorized as under 100 employees) were using cloud computing. But with the ever-growing selection of premier providers and regional options, it would appear that this number is poised to see dramatic growth.

In our own practice in the Philadelphia area, we are leveraging cloud-based services in every area of what we do. For a small customer with limited budget or a new startup without huge amounts of seed capital, we can steer them into the Microsoft Business Productivity Online Suite, or BPOS. This recent Microsoft offering provides anything from simple e-mail hosting right up through hosted Instant Messaging and SharePoint site. And with mobility being paramount for many, these solutions can be integrated with the smartphone of choice to make it accessible from anywhere.

In some cases, a more traditional server/client approach is needed, and we provide this either through an onsite implementation at the customer location, or completely cloud-oriented with virtual servers in our hardened data center.[6] We have revolutionized this concept with our flagship Managed Environment program. Provided on a per-user basis is all the hardware, software and labor services necessary, based upon the custom solution. This allows the client to avoid a substantial capital expenditure, create a tax-benefitting operating expense, and cancel the need for expensive full-time IT staff. There are firms that have scrapped their aging network infrastructure and moved lock, stock and barrel to this model. In many cases, they've added services that didn't make sense with their old solutions, including internal instant messaging and presence, hosted VoIP telephony, and multi-layered backups and disaster recovery. Essentially, the small business now has access to enterprise-class technology and security without the astronomical investment.

Whether it is specific cloud services, or a full-blown all-encompassing solution, it is apparent that this revolution is just in its nascent stages. Despite recent failures in the marketplace, like Amazon's days-long outage, it would seem clear that the direction of both small and large organizations is clearly headed up in the cloud. The potential for cost savings, productivity enhancements, and increased employee mobility and satisfaction will drive the providers to even greater levels of reliability. And even though the Jetsons "air cars" are a long way from reality, it would seem in the short term that the future of IT network infrastructure lies clearly in the air above. Who knew? What's even more amazing is that I didn't even need Accu-Weather to predict it!

New Words and Expressions

tenuous　*adj.*　纤细的，稀薄的，贫乏的
ENIAC（electronic numerical integrator and calculator）　电子数字积分计算机
fill-in　*n.*　代替者，补充物　*adj.*　临时填补的
Cape Cod　科德角（美国地名）
namesake　*n.*　名义，同名物，同名的人
postulate　*v.*　假定，要求　*n.*　基本条件，假定
decentralization　*n.*　分散，非集权化，疏散
latency　*n.*　潜伏，潜在因素
traction　*n.*　牵引，牵引力
private cloud　私有云
resale　*n.*　转售，零售，再贩卖
public cloud　公有云
cloud computing　云计算
grid computing　网络计算
Software-as-a-Service（SaaS）　软件即服务
Infrastructure-as-a-Service（IaaS）　基础设施即服务
Platform-as-a-Service（PaaS）　平台即服务
inventory　*n.*　存货，存货清单，详细目录，财产清册
interoperability　*n.*　互操作性，互用性
portability　*n.*　可移植性，轻便，可携带性
tout　*v.*　推销，兜售
leapfrog　*v.*　跳背，跳蛙，交替前进，跃过
leverage　*v.*　杠杆式投机，举债经营

host environment　宿主环境
mediocre　*adj.*　普通的，平凡的，中等的
offsite　*n.*　装置外，厂区外
trade-off　*n.*　交换，交易，权衡，协定
SMB（Small & Medium Business）　中小企业
onsite　*adj.*　在场的，就地的
nascent　*adj.*　初期的，开始存在的，发生中的
headed up　领导，抬高
Philadelphia　*n.*　费城（美国宾夕法尼亚州东南部港市）
VoIP（Voice over Internet Portocol）　互联网协议电话

Notes

[1] More than two decades later, I actually did get the opportunity to be a TV weatherman, though just as a fill-in for the vacationing regular on a tiny station on Cape Cod, Massachusetts in my former life as a sports reporter. 20多年以后，我确实有机会成了电视天气预报员，尽管那只是我前半生当体育播报员时在马萨诸塞州科德角一个小电台做临时假期的替补工作者。

[2] And what are the good and bad consequences of adopting this growing trend to your business, especially with the tenuous economy and the more pronounced effects on the small to midsized business. 还有这种持续的增长趋势会给商业带来什么样的影响呢？特别是对于中小型商业的薄弱经济和显著影响。

[3] Early machines relied on faster internal workings to do this, until the trend became combining many different machines in a cluster on one site to maximize the load sharing. 早期的计算机依靠更快速的内部工作元件来实现这一功能，直到将许多不同的计算机在一个站点组合成群以实现最大化的资源共享成为趋势，这一现象才得以改变。

[4] The ever-growing availability of higher-speed connectivity, along with advances in data storage and increased input/output efficiency has allowed this previous notion of grid computing to evolve. 逐渐被采用的高速联接方式，随着数据存储技术的发展和输入输出效率的不断提高，已经促使了网络计算原有概念的进一步演化。

[5] Taking many, or even thousands, of servers out of service, replaced by "virtual" servers in a larger host environment is becoming more and more the standard by which data centers are being built or rebuilt. 将许多甚至成千上万的服务器退役，并用大规模宿主环境下的虚拟服务器替代，这正在逐渐成为数据中心建设和重建的标准。

[6] In some cases, a more traditional server/client approach is needed, and we provide this either through an onsite implementation at the customer location, or completely cloud-oriented with virtual servers in our hardened data center. 在许多情况下，需要一种传统的服务器/客户机的模式，我们可以在用户所在位置就地实现，或者用可靠的数据中心内的虚拟服务器完全借助云的指导来实现。

C Challenges and Benefits of Industry 4.0（工业4.0的挑战和优势）

1. Challengesof Industry 4.0

Not every observer is convinced of the value that Industry 4.0 will add. Some feel that Industry

4.0 as a concept is poorly defined and suffers from exaggerated expectations; others believe that fully digitised products and value chains are still a "pipe dream". [1] The Gartner Group's hype cycle for emerging technologies for 2014 places many of the technologies associated with Industry 4.0 [including machine-to-machine communications, big data, the Internet of things (IoT) and smart robots] near the "peak of inflated expectations", still five to ten years from the point where the payoff for applying these in the broad market is evident. In a global 2013 – 2014 survey, 88% of the respondents said they did not fully understand the underlying business models of the industrial Internet of things and its long-term implications for their industry. Even those convinced of the value of Industry 4.0 can foresee a series of barriers ahead.

(1) Investment and Change

Building a complex value network that can produce and distribute products in a flexible fashion means business leaders must accept to change and partner with other companies—not only suppliers and distributors of a product, but technology companies and infrastructure suppliers such as telecoms and Internet service providers. [2] Companies may even need to cooperate with competitors, e.g. in the establishment and use of standards that allow the transmission and exploitation of large quantities of data. Large investments are needed if enterprises are to make the move to Industry 4.0; these are projected to be €40 billion annually until 2020 for Germany alone (perhaps as much as €140 billion annually in Europe). These investments can be particularly daunting for small and medium-sized enterprises (SMEs) who fear the transition to digital because they cannot access how it will affect their value chains. [3] So far take up has been cautious: Even in Germany (a leader in manufacturing), only an estimated one in five companies uses interconnected IT systems to control its production processes, though almost half intend to do so. Some critics say that systems are too expensive, too unreliable and oversized, and that the Industry 4.0 approach is being driven largely by equipment producers rather than customer demand.

(2) Data Ownership and Security

With the large quantities of data being collected and shared with partners in the value network, businesses need to be clear about who owns what industrial data and to be confident that the data they produce will not be used by competitors or collaborators in ways that they do not approve. [4] In particular, smart services will be based on the data generated by smart devices during their manufacture and use. For example, car-makers are reluctant to share data generated by their cars, for fear of finding their profits being squeezed by digital competitors. A single set of European rules on privacy, data storage and copyright, that balances trust and data protection, is considered by some to be a necessary step to ensuring European competitiveness.

(3) Legal Issues

Advanced manufacturing also raises a variety of legal questions including employee supervision, product liability and intellectual property. For example, data from a "smart glove" that guides and records the movements of workers might be used to monitor or evaluate employees. If an autonomous manufacturing system that links different value networks produces a defective or dangerous product, how should the courts determine who in the network is responsible? If a customer requests an individualised product, who owns the intellectual property (IP) rights to the design? The French Conseil "d'analyse économique" has called for a balance between the stimulation of innovation by protecting IP and the sharing of knowledge, both sources of future progress.

(4) Standards

Standards are essential to ensure the exchange of data between machines, systems and software within a networked value chain, as a product moves into and through the "smart factory" towards completion, as well as to allow robots to be integrated into a manufacturing process through simple "plug-and-play" techniques. If data and communication protocols are proprietary or only recognised nationally, only the equipment of one company or group of companies will be compatible; competition and trade can be expected to suffer and costs rise. On the other hand, independent, commonly agreed, international standard communication protocols, data formats and interfaces can ensure interoperability across different sectors and different countries, encourage the wide adoption of Industry 4.0 technologies, and ensure open markets worldwide for European manufacturers and products. A foresight study by the Joint Research Centre emphasised the need for anticipating standards requirements and accelerating their development in Europe.

(5) Employment and Skills Development

The nature of manufacturing work has been shifting from largely manual labour to programming and control of high performance machines. Employees with low skill levels risk becoming replaceable unless they are retrained. On the other hand, workers able to make the transition to Industry 4.0 may find greater autonomy and more interesting or less arduous work. Employers need personnel with creativity and decision-making skills as well as technical and ICT expertise. By 2020, labour markets in the EU could be short of as much as 825,000 ICT professionals; this shortage may be even more pronounced in advanced manufacturing settings where big data analysts and cybersecurity experts are required. While various initiatives have been undertaken to encourage the acquisition of "eSkills", young people may not necessarily be interested by the digitalisation of the workplace: In one survey only 13% of young adults in Germany would definitely consider a career in ICT despite the majority view that the sector offered the best job prospects.

2. Benefits of Industry 4.0

Whether it's Industry 4.0, smart industry or the industrial Internet, there are ample benefits for manufacturers to transform the way they work. We've mentioned some benefits, risks and challenges earlier in this overview but let's look a bit closer at some of the main advantages.

The essential goal of Industry 4.0 is to make manufacturing—and related industries such as logistics—faster, more efficient and more customer-centric, while at the same time going beyond automation and optimization and detect new business opportunities and models.[5]

Most of the benefits of Industry 4.0 are—obviously—similar to the benefits of the digital transformation of manufacturing, the usage of the IoT in manufacturing, operational and business process optimization, information-powered ecosystems of value, digital transformation overall, the industrial Internet and many other topics on our website. However, let's summarize a few of the key benefits of Industry 4.0.

(1) Enhanced Productivity through Optimization and Automation

As mentioned in the section on the state of Industry 4.0, optimization of processes and of productivity is the first benefit that manufacturers see. It's also one of the first goals of Industry 4.0 projects. In other words: Saving costs, increasing profitability, reducing waste, automating to prevent errors and delays, speeding up production to work more in real-time and in function of the overall value chain, where speed is crucial for everyone, digitizing paper-based flows, being able to intervene faster in case of production issues and so forth. It's the low hanging fruit, yet important. On top of the research from BCG (Boston Consulting Group), the signs that investments are done in these areas first are clear. Again, it's not a coincidence that, from a spending perspective, the number one use case in which manufacturers invest their IoT budgets is manufacturing operations (a whopping $102.5 billion on a total of IoT $178 billion across all manufacturing use cases in 2016). Industry 4.0 offers various solutions to optimize, from optimized asset utilization and smoother production processes to better logistics and inventory management.

(2) Real-Time Data for a Real-Time Supply Chain in a Real-Time Economy

While we just mentioned speed in a context of optimization, automation and enhanced productivity, it is a benefit in many other ways as well. A lot of the productivity improvement benefits are rather about the internal goals of costs and process optimization. Yet, at the same time several also fit in a perspective of enhanced customer-centricity. Industry 4.0 is about the entire life cycle of products and manufacturing obviously doesn't stand on its own. If you look at the entire value chain and ecosystem within which manufacturing operations reside there are many stakeholders involved. These are all customers. And customers also want enhanced productivity, regardless of where they sit in the supply chain. If the final customer wants good products fast and has increased expectations regarding customer experience, quality, service and products that are delivered on the exact time they want, this impacts the whole supply chain, all the way up to manufacturing and beyond. Speed is not just a competitive advantage and customer expectation in an increasingly real-time economy, it's also a matter of alignment, costs and value creation. Moreover, customers simply expect it. Once again the crucial role of data and information surfaces. Industry 4.0, smart factories, supply chains, informed customers, alignment: it's all about data, from the actual operations to the delivery of a product to an end customer and beyond. The more data you gather early on and the more timely this data gets where it matters when it matters, the more value down the supply chain. In fact, this is the essence of one of the three dimensions of RAMI 4.0, the Reference Architecture Model Industry 4.0, which we tackle below.

(3) Higher Business Continuity through Advanced Maintenance and Monitoring Possibilities

When an industrial asset gets broken it needs to be fixed. That costs time, money and very often a lot of moving around by support people and engineers. When a key industrial asset, such as an industrial robot in a car manufacturing plant gives up, it's not just the robot that's broken. Production is affected, costing loads of money and unhappy customers, and sometimes production can be fully disrupted. It's everyone's worst nightmare as business continuity is an extremely high concern. On top of all the replacement/fixing work, resources and costs, reputation can be damaged, orders can be cancelled and with each hour that passes money is thrown away. If industrial assets are connected and can be monitored (health status monitoring, for instance) through the IoT and issues are tackled before they even happen the benefits are huge. Alerts can be set up, assets can be proactively maintained, real-time monitoring and diagnosis becomes possible, engineers can fix issues, if they do occur from a distance, the list goes on. [6] Moreover, patterns and insights are gained to optimize in areas where things seem to have issues more often and a world of new maintenance services opens up as we'll see. No wonder that asset management and maintenance are the second largest area of IoT investments in manufacturing.

(4) Better Quality Products: Real-Time Monitoring, IoT-Enabled Quality Improvement and Cobots

We mentioned that customers want speed. However, that doesn't mean they are ready trade quality for speed, well on the contrary. If you have everything in your production system and its broader environment hooked up with sensors, software, IoT technologies, systems of insight and the customer, you can also enhance quality of your products. Automation definitely plays a big role here and so do the typical components of cyber-physical systems (more below) and the IoT whereby quality aspects can be monitored in real-time and robots reduce errors. [7] On the flip side and one of the risks and challenges to tackle, as mentioned earlier: The more you automate, the less work for people, in theory. And the same goes for other mentioned benefits such as maintenance (the less you need engineers for support, the less support engineers you need). It's a dilemma and known issue which we'll cover later. In the meantime do know that robots are not going to take all human jobs over soon. Ample companies have increased the usage of robots and at the same time hired more. The reason we mention it in the context of quality is that this is certainly one area where you see cobots popping up ("cobots" is a fancy term for advanced collaborative robots or put more simply: robots that fit a collaboration between man and machine).

(5) Better Working Conditions and Sustainability

Talking about people, the human (and social) dimension is ubiquitous in Industry 4.0. Moreover, if we look at the possibilities and benefits, that human, social and even environmental aspect is key in the goals of Industry 4.0. Improving working conditions based on real-time temperature, humidity and other data in the plant or warehouse, quick detection and enhanced protection in case of incidents, detection of presence of gasses, radiation and so forth, better communication and collaboration possibilities, a focus on ergonomics, clean air and clean factory initiatives (certainly in Industry 4.0 as the EU wants to be leading in clean air and clean anything technologies), the list goes on.

(6) Personalization and Customization for the "New" Consumer

We all know it: Consumer behavior and preferences have changed. Digital tools have changed the ways we work, shop and live. People have also become more demanding, among others with regards to fast responses and timely information/deliveries as mentioned earlier. On top of that consumers also like a degree of personalization, depending on the context. Take sports shoes, for instance. Once a few colors of the same shoe were enough, know we want the ability to customize them in whatever way. On top of that another phenomenon is taking place and it does disrupt traditional supply chains. Consumers increasingly get (and want) possibilities to have a direct interaction with a brand and its manufacturing capability. Digital platforms to customize products as mentioned, shortened routes between production and delivery, possibilities to co-create and so on. In many manufacturing environments these things already happen. And it's not just in a consumer environment. We increasingly see customization in a B2B (business-to-business) context as well, even if it's just to stick a label, add a custom feature or adapt any characteristic of the product whatsoever. If you want to offer these services at scale and even turn them into a competitive advantage, automation and several technologies and processes in industry 4.0 become a necessity. A real-life example without disclosing the details: A large bank wanting specific office equipment to use across all its branches (customer-facing context) with its own look, feel and features as part of a rebranding. There are plenty more examples.

(7) Improved Agility

Now that we speak about competitive benefits and customization we also need to tackle agility, scalability and flexibility. The same scalability and agility which we expect from supporting IT services and technologies, such as the cloud, are expected in manufacturing. This is partially related with the previous topic of customization but mainly is about leveraging technologies, Big Data, AI, robots and cyber-physical systems to predict and meet seasonal demand, fluctuations in production, the possibility to downscale or upscale. In other words, all the adjustments that are sometimes more or less predictable, can be made more predictable or are not predictable but can be handled thanks to increased visibility, flexibility and a possibility to leverage assets in function of optimal production requirements from a perspective of time and scale.[8]

(8) The Development of Innovative Capabilities and New Revenue Models

Digital transformation, as you can read in our digital transformation strategy overview, is a matter of many levels, steps and capabilities. You can transform processes, specific functions, customer service, experiences and skillsets but in the end true value is generated by tapping into new, often information-intensive, revenue sources and ecosystems, enabling innovative capabilities, for instance in deploying an as-a-service-capacity for customers, advanced maintenance services and so on.

New Words and Expressions

pipe dream　脱离实际的愿望，行不通的计划，妄想
daunt　v.　使胆怯，恐吓，使畏缩
ample　adj.　足够的，丰裕的
low hanging fruit　唾手可得的成果
BCG (Boston Consulting Group)　是一家全球性管理咨询公司，是世界领先的商业战略咨询机构，客户遍及所有行业和地区

alignment　n.　排成直线，（国家、团体间的）结盟
disrupt　v.　扰乱，打乱，使中断
B2B (business-to-business)　企业对企业
cobot　n.　合作机器人，是在同一作业空间内直接与人合作的机器人
dilemma　n.　困境，左右为难，进退两难的局面

Notes

[1] Some feel that Industry 4.0 as a concept is poorly defined and suffers from exaggerated expectations; others believe that fully digitised products and value chains are still a "pipe dream". 有些人认为工业4.0是个简单定义的夸大其词的概念；另一些人认为全数字化产品和价值链仍然是个不切实际的愿望。

[2] Building a complex value network that can produce and distribute products in a flexible fashion means business leaders must accept to change and partner with other companies—not only suppliers and distributors of a product, but technology companies and infrastructure suppliers such as telecoms and Internet service providers. 建立一个能够以灵活的形式生产和分配产品的复杂价值网络就意味着商业领导必须接受变化和其他公司的合作者——不仅是产品的供应商和经销商，还有技术公司以及如电信和网络服务的基础供应商。

[3] These investments can be particularly daunting for small and medium-sized enterprises (SMEs) who fear the transition to digital because they cannot access how it will affect their value chains. 这些投资者会受到中小型企业的威胁，而这些企业会因无法获取如何影响价值链的信息而害怕数字化转变。

[4] With the large quantities of data being collected and shared with partners in the value network, businesses need to be clear about who owns what industrial data and to be confident that the data they produce will not be used by competitors or collaborators in ways that they do not approve. 在价值网络中随着收集和合作者共享数据量的增大，商家需要知道谁拥有工业数据，并且确信他们的生产数据不会在没有允许的情况下被竞争者或者合作者使用。

[5] The essential goal of Industry 4.0 is to make manufacturing—and related industries such as logistics—faster, more efficient and more customer-centric, while at the same time going beyond automation and optimization and detect new business opportunities and models. 工业4.0的核心目标是使制造和相关的工业，如物流——更快、更有效和更加以客户为中心，一次同时实现自动化、优化以及新商业机会和模式的探测。

[6] Alerts can be set up, assets can be proactively maintained, real-time monitoring and diagnosis becomes possible, engineers can fix issues, if they do occur from a distance, the list goes on. 建立警报、提前保护资产、实时监视和诊断都会成为可能，如果这些情况是远距离发生的，工程师们都可以处理，而这样的事例还在不断增加。

[7] Automation definitely plays a big role here and so do the typical components of cyber-physical systems (more below) and the IoT whereby quality aspects can be monitored in real-time and robots reduce errors. 自动化在这里确实非常重要，更底层的信息物理系统的典型组件也很重要，并且由此能够实时监视物联网的质量，用机器人减少误差。

[8] This is partially related with the previous topic of customization but mainly is about leveraging technologies, Big Data, AI, robots and cyber-physical systems to predict and meet seasonal demand, fluctuations in production, the possibility to downscale or upscale. In other words, all the adjustments that are sometimes more or less predictable, can be made more predictable or

are not predictable but can be handled thanks to increased visibility, flexibility and a possibility to leverage assets in function of optimal production requirements from a perspective of time and scale. 这与前面的定制主题部分相关，但主要是关于关键技术，用大数据、人工智能、机器人和信息物理系统去预测，以满足季度需求、产量浮动、规模扩大或者缩小的可能性。换句话说，由于从时间和规模上优化了生产，增加了可视性、灵活性和资产的可操控性，所有的、或多或少可以预测的调节会变得更好预测，或者尽管不能预测但却可控。

PART II Practical Writing Techniques
（实用的写作技巧）

Unit 1 Abstracts（摘要）

摘要是科技论文（研究报告、期刊文章、毕业论文、会议论文）的重要组成部分。它以提供文献内容梗概为目的，不加评论和补充解释，简明、确切地记述文献重要内容。

随着国际交流的日益广泛，科技论文一般都要求同时附有英文摘要。如今，英文摘要成了国家间进行知识传播、学术交流和合作的媒介。然而，调查显示，目前科技论文中的英文摘要绝大多数比较粗糙，不能很好地满足国际交流的要求。

本单元将在简单介绍摘要基本类型的基础上，阐述摘要的写作要素以及摘要撰写技巧，最后给出摘要各部分的基本句型。

1. Types of Abstracts（摘要的类型）

根据原文献的类型和形式，摘要可分为三种类型，即报道性摘要（informative abstract）、指示性摘要（indicative or descriptive abstract）和报道-指示性摘要（informative-indicative abstract）。一般文摘杂志、数据库和期刊论文所附摘要都属于这三类摘要。由于报道-指示性摘要接近于报道性摘要，所以通常也将摘要分为报道性摘要与指示性摘要两大类。

（1）报道性摘要

报道性摘要即资料性摘要或情报性摘要。报道性摘要比较详尽，强调尽可能完整地报道原文献的具体内容，提供尽可能多的定量或定性的情报信息，基本上反映出科研成果的全部创造性内涵。这类摘要的作用是帮助读者对是否查阅原文献做出判断，以及输入计算机长期储存作为情报服务的重要内容，起到代替原文献的作用。因此这类摘要适用于研究或试验报告、研究论文及专题论文，其篇幅稍长，一般研究论文或试验报告的摘要为 100~250 个英文单词。内容很多很充实的文献，如博士学位论文或长篇研究报告等的摘要可达 500 个英文单词。

（2）指示性摘要

指示性摘要即概述性摘要或简介性摘要。它只简要地介绍论文的论题，或者概括地表述研究的目的，仅使读者对论文的主要内容有一个概括的了解。篇幅以 50~100 个英文单词为宜。

指示性摘要适用于综述性、讨论性文章、简短论文以及普通报道等一类文体。它仅向读者介绍文献的一般内容，如研究的目的、范围等，不涉及文献的具体内容，如研究的方法、结果和讨论。这类摘要篇幅较短，有时是一句话。

（3）报道-指示性摘要

报道-指示性摘要是以报道性摘要的形式表述论文中价值最高的那部分内容，其余部分则以指示性摘要形式表达。篇幅以 100~200 个英文单词为宜。

报道-指示性摘要是指既有报道性又有指示性的摘要，是一种在科技刊物上比较多见的摘要形式。由于论文摘要篇幅的限制，常常需要把论文的主要方面写成报道性而次要方面写成指示性，这样就得到报道–指示性摘要。这种摘要比前两种类型的摘要更能达到以最短的篇幅传递最大信息量的目的。因而在期刊论文中用得更为普遍，效果也较好。

一般的科技论文都应尽量写成报道性摘要，而对综述性、资料性或评论性的文章可写成指示性或报道–指示性摘要。

2. Essential Elements of the Abstract（摘要的基本要素）

文章摘要实质上是对论文的高度浓缩，因此其基本结构和内容与论文对应。从结构上来讲，文章摘要分为研究目的、方法、结果和结论几个部分。

1）目的部分主要阐述研究工作的范围、背景、目的、需解决问题以及意义。

2）方法部分主要阐述采用什么方法解决了研究目的部分提出的问题。

3）结果部分主要是实验的观测结果和数据，一般给出比较重要的结果。

4）结论部分一般是对实验结果进行分析或评价，也可以对结果的意义进行简单说明（是否有意义、有何重要意义或者毫无意义等）。

3. Writing Techniques of the Abstract（摘要撰写技巧）

一个好的摘要应该具有以下几个特点：

1）准确——一个好的摘要仅仅包含原始文章中的信息。

2）简洁——一个好的摘要应该切中要点，语言精确，不包含多余的形容词。

3）清晰——一个好的摘要不包含术语或者口语，并对缩写进行解释。

4）完整——一个好的摘要应包含事情的完整信息。

撰写摘要的工作应该在论文完成后进行。首先要通读全文，在通读过程中要记得自己的任务是书写摘要，因此重点关注文章的目的、方法、结果和结论等关键部分。通读完后就可以着手进行撰写。撰写摘要时应注意以下几点：

（1）内容与写作特点

摘要的内容应包括原文的主题内容以及定性定量的结果。要文笔精炼、重点突出、结果明确，逻辑性强；内容要准确具体，避免笼统抽象模糊不清的叙述。不能写入原文没有的内容，也不能单纯照抄论文中的关键句子，更不能出现内容及文字表达方面的错误。

（2）独立完整性

写作摘要的目的是使读者在阅读原文之前能了解原文的主要内容，以便决定是否需要阅读原文。特别是在文摘刊物或某些会议的论文集中，摘要是单独发表的。因此，摘要应如同一篇精炼的短文，具有独立完整的结构和内容，使读者仅仅通过摘要即可获知原文的主要信息。

（3）段落与篇幅

摘要要求简短，通常期刊论文的摘要不分段（也有分段的）。长摘要可分段，如学位论文摘要等。摘要的篇幅视原文的情况决定。根据国际标准化组织的规定，对于大多数试验、研究论文和一部分专著，摘要长度为 250 个英文单词就可以了；学位论文的摘要不超过 500 个英文单词。根据美国多数高等院校规定，硕士论文摘要以 250 个英文单词左右，博士论文摘要以 350 个英文单词左右为宜。一般发表在英文学术期刊上的论文摘要，其篇幅为英文长度的 1%～3%。

（4）用词造句

摘要中必须使用正规英语和标准术语，一般不用缩写形式，不应包含公式、表格、化学式、图表、参考文献标记，以及人们不熟悉的术语、词汇、缩略词、符号等。需要使用缩略词或符号时，应在第一次出现时加以说明，以保持摘要的完整性和独立可读性。摘要中应使用完整的句子，不用电报式的简略文字，可使用转折词和短语使摘要前后连贯。

（5）人称

摘要一般以第三人称语气或非人称语气书写，少用第一人称，以反映科技论文的客观性。因此在摘要中一般不用"我""我们"等字样。随着科技英语的发展，近来国外科技论文及摘要中使用第一人称的情况有所增加，但是使用第三人称语气的被动句依然是描述客观科学技术和方法的主流形式。

（6）时态与语态

摘要的时态大都使用一般现在时或一般过去时，少量采用现在完成时。要注意摘要中时态的一致性以及与正文时态的相对一致性。国外科技刊物论文摘要中使用的谓语动词，仍以被动语态居多，这主要是为了表达内容的客观性。不过主动语态可使句子清楚、简洁、有力，目前有增多应用的趋势。以下是写科技文献摘要的时态常用的方法：

1）介绍背景资料时，如果句子的内容为不受时间影响的普遍事实，应使用现在时；如果句子的内容是对某种研究趋势的概述，则使用现在完成时。

2）在叙述研究目的或主要研究活动时，如果采用"论文导向"，多使用现在时（如 This paper presents...）；如果采用"研究导向"，则使用过去时（如 This study investigated...）。

3）概述实验程序、方法和主要结果时，通常用现在时。实验过程部分一般采用一般过去时。

4）叙述结论或建议时，可使用现在时、臆测动词或 may, should, could 等助动词。

（7）非英语人名与地名

摘要中出现俄罗斯、德国、法国、日本等其他国家的人名时，应将它译成英文，不要将几种语言文字夹杂使用；音译名除个别沿用已有者外，均应按照汉语拼音方案的拼法译出；中国的地名、人名均用汉语拼音字母书写。

（8）标题

论文的标题通常可视为摘要的一部分，摘要的首句须避免重复标题；如果论文标题的概括性不够，可在摘要的首句做适当的补充。在标题之后依次列出作者姓名、发表论文的期刊名称、卷数、期数、起讫页数、出版年代等。

4. Key Words（关键词）

除了摘要外，有时候还要求学术论文附关键词。关键词是科技论文的文献检索标志，是表达文献主题概念的自然语言词汇，它是为了适应计算机检索的需要而提出来的，位置在摘要之后。

国家标准对于关键词的定义是：关键词是为了文献标引工作从报告、论文中选取出来的用以表示全文主题内容信息的单词或术语。由此可见，学术论文的关键词必须是单词或术语。如多传感器集成、数据融合、程序性能预测等可以作为关键词。在科学技术信息迅猛发展的今天，全世界每天有数万篇科技论文呈献，关键词成为检索最新发表论文的手段，关键词选得是否恰当，关系到该文被检索和该成果的利用率。

关键词包括叙词和自由词。叙词是指收入《汉语主题词表》《MeSH》（Medical Subject Heading Terms，医学主题词）等词表中可用于标引文献主题概念的即经过规范化的词或词组。自由词是反映该论文主题中新技术、新学科尚未被主题词表收录的新产生的名词术语或在叙词表中找不到的词。为适应计算机自动检索的需要，GB/T 3179—1992 规定，现代科技期刊都应在学术论文的摘要后面给出 3~8 个关键词（或叙词），各关键词以逗号或者分号分开。关键词的标引应按 GB/T 3860—1995《文献叙词标引规则》的原则和方法，参照各种词表和工具书选取；未被词表收录的新学科、新技术中的重要术语以及文章题名的人名、地名也可作为关键词（自由词）标出。

提炼出科技论文关键词的基本方法是：首先对文献进行主题分析，弄清该文的主题概念和中心内容；其次，尽可能从题名、摘要、层次标题和正文的重要段落中抽出与主题概念一致的词和词组；再次，对抽出的词和词组进行重要性排序，从中选出最重要的 3~8 个；最后，根据叙词表对选出的词和词组进行规范化或者把叙词组配成专指主题概念的词组，还有部分无法规范为叙词的词，但是表达主题概念所必需的，可作为自由词标引并列入关键词。

5. Sample Abstracts（摘要举例）

（1）A Sample of Informative Abstract（报道性摘要举例）

报道性摘要需要全面、简要地概括论文的目的、方法、主要数据和结论。常用的写作形式如下例：

Sample 1

Design & Implementation of Smart House Control Using LabVIEW

Smart home is a house that uses information technology to monitor the environment, control the electric appliance and communicates with the outer world. Smart home is a complex technology, at the same time it is developing. A smart home automation system has been developed to automatically achieve some activities performed frequently in daily life to obtain more comfortable and easier life environment. A sample house environment monitor and control system that is one branch of the smart home is addressed in this paper. The system is based on the LabVIEW software and can act as a security guard of the home. The system can monitor the temperature, humidity, lighting, fire & burglar alarm, gas density of the house and have infrared sensor to guarantees the family security. The system also has Internet connection to monitor and control the house equipment's from anywhere in the world. This paper presents the hardware implementation of a multiplatform control system for house automation using LabVIEW. Such a system belongs to a domain usually named smart house systems. The approach combines hardware and software technologies. Test results of the system have shown that it can be easily used for the smart home automation applications.

Key words: smart house; LabVIEW; PIC16F877A; data acquisition card; remote control

第二种形式的报道性摘要又称为结构式摘要，其文中用特殊字体标明目的、方法、结果和结论，可以让读者一目了然。例如：

Sample 2

Models of Information Exchange for UK Telehealth Systems

O. Adeogun, A. Tiwari, J. R. Alcock

Aim: The aim of the paper was to identify the models of information exchange for UK telehealth systems.

Methodology: Twelve telehealth offerings were evaluated and models representing the information exchange routes were constructed. Questionnaires were used to validate the diagrammatical representations of the models with a response rate of 55%.

Results: The models were classified as possessing four sections: preparing for data transfer, data transfer, information generation and information transfer from health professional to patient. In preparing for data transfer, basic data entry was automated in most systems though additional inputs (i. e. information about diet, lifestyle and medication) could be entered before the data was sent into the telehealth system. For the data transfer aspect, results and additional inputs were sent to intermediate devices, which were connectors between point-of-care devices, patients and health professionals. Dataware then forwarded to either a web portal, a remote database or a monitoring/call centre. Information generation was either through computational methods or through the expertise of health professionals. Information transfer to the patient occurred in four forms: e-mail, telehealth monitor message, text message or phone call.

Conclusion: On comparing the models, three generic models were outlined. Five different forms of information exchange between users of the system were identified: patientpush, system-stimulation, dialogue, health professional-pull and observation. Patient-push and health professional-pull are the dominant themes from the telehealth offerings evaluated.

Key words: information exchange; model point-of-care device; telehealth

（2）A Sample of Indicative Abstract（指示性摘要举例）

指示性摘要一般只用两三句话概括全文，不涉及结论和论据，多用于综述和会议报告等，如下面的中文综述摘要可以直接译为指示性摘要。

智能控制综述

中文：目前智能控制已经广泛应用于各个工业领域，本文主要阐述了智能控制的基本概念和分类，并详细介绍了应用于智能控制的神经网络（neural network）、模糊逻辑（fuzzy logic）、进化计算（evolutionary computation）等人工智能方法，最后为科研人员指明了未来智能控制的发展方向。

关键词：智能控制；神经网络；模糊逻辑；进化计算

英文：Nowadays intelligent control has been widely used in various industrial fields. Basic concepts and categories of intelligent control are firstly narrated in this paper. Furthermore, some artificial intelligence approaches applied to intelligent control, such as neural network, fuzzy logic and evolutionary computation, have also been introduced thoroughly. Future developing directions of intelligent control are finally presented for researchers.

Key words: Intelligent control; Neural network; Fuzzy logic; Evolutionary computation

6. Useful Sentence Structures in English Abstracts（英文摘要常用句式）

（1）开头部分

开头部分一般用来回顾研究背景，可以是阐明写作或研究目的，也可以是介绍论文的重点内容或研究范围，其常用句型有：

The goal of this paper is…

本论文的目标是……

The primary goal of this research is...
本研究的主要目标是……
The intention of this paper is to...
本论文的目的是……
The overall objective of this study is...
本研究的总体目标是……
The chief aim of the present work is to...
当前工作的主要目的是……
We present a...
我们提出一种……
This report describes...
这篇报道描述了……
This article mainly discussed...
本文主要讨论……
A study... is reported...
报道了……研究

例如：

The goal of this paper is to show that they are also highly useful for designing shock capturing numerical schemes for hyperbolic conservation laws.

We present a variational framework for the registration of tensor-valued images.

We present a new denoising method for DTI with low SNR and explore it by Monte Carlosimulations based on a phantom and on human brain data with high resolution 1 mm^3 voxels.

This report describes the characteristics of both Venus and Mercury, covering size, rotation, revolution about the sun, density, physical appearance and atmosphere.

This article mainly discusses neural networks and their application.

A study of the hydrodynamics of drafting, initiated in the hope of understanding the mechanisms causing the separation of mothers and calves during fishing-related activities, is reported here.

（2）方法介绍部分
该部分一般是介绍试验过程，试验方法或者应用、用途等。常用句型有：
The method used in our study is known as...
我们研究中使用的方法就是所谓的……
The approach adopted extensively is called...
这个广泛被采用的方法叫作……
The procedure they followed can be briefly described as...
他们采用的步骤可以简单描述为……
Detailed information has been acquired by the authors using...
作者采用……获得了详细信息。
The fundamental feature of this theory is as follows...
该理论的基本特征如下……

The experiment consisted of three steps, which are described in...
该实验包括三个步骤,其在……进行了描述。
We propose...
我们提出……
The proposed...
提出的……
This formula is verified by...
该公式通过……得到验证。
例如:
The proposed extension allows to efficiently integrate, in the deconvolution process, a set of soft constraints given by a probabilistic MRI atlas containing experts' prior knowledge about the spatial localization of the different brain structures.

We propose and investigate a novel flux corrected transport (FCT) type algorithm. It is composed of an advection step capturing the flow dynamics, and a stabilized nonlinear backward diffusion step in order to improve the resolution properties of the scheme.

The method used in our study is known as Hamilton function.

The fundamental feature of this theory is as follows. Boundary conditions are simulated through equivalent stiffness and mass. The cable structure is studied to form a dynamical model considering support vibration.

(3) 结论部分
结论部分常用句型有:
In summing up it may be stated that...
总之,它可以描述为……
It is concluded that...
结论是……
The results of the experiment indicate that...
实验结果表明……
The studies we performed showed that...
我们进行的研究表明……
The investigation carried out by... has revealed that...
……进行的调查表明了……
Laboratory studies of... did not furnish any information about...
对于……的实验室研究没有给出关于……的任何信息
This fruitful work...
这项卓有成效的工作……
These findings of the research have led the author to the conclusion that...
这项研究得到的这些结果使得作者得出……的结论
Finally, a summary is given of...
最后,对于……给出了一个结论。

例如：

In summing up it may be stated that a toric surface can be used to describe the form deviation of optical elements.

Theresults of the experiment indicate that the performance control system can adjust the network configuration dynamically to improve the network performance on the trigger of the performance event according to the performance control policy.

Unit 2　Resumes（简历）

1. Introduction（概述）

简历，顾名思义，就是对个人学历、经历、特长、爱好及其他有关情况所做的简明扼要的书面介绍。简历是个人形象，包括资历与能力的书面表述，对于求职和求学者而言，是获得一份工作或者进入一个学校的敲门砖。

简历的英文为 resume，该词来源于法语，与其意义相近的还有 personal history、personal record data、data sheet 等，但 resume 最常用。本单元将介绍简历的主要内容和写作。

2. Writing Techniques of the Resume（简历写作技巧）

个人简历可以是表格的形式，也可以是其他形式。完整的个人简历主要包括如下几个部分：

（1）个人资料（personal data）

个人资料包括个人姓名、性别、出生年月、家庭地址、政治面貌、婚姻状况、身体状况、兴趣、爱好、性格、联系电话等。地址和电话等联系信息是用人单位第一时间找到你最重要的信息，以便必要时和你取得联系。其他个人信息，如会乐器或长期坚持某项体育活动等兴趣爱好，可以表明自己健康阳光的形象，表现自己综合的素养。如果版面有限可以根据具体情况简略或者省去。

（2）求职目标（objective/position wanted/position sought）

对于职业目标比较明确的人来讲，在简历中突出自己的求职目标是非常有好处的。一般将职业目标放到名字和联系方式下面。通常典型的求职目标以助动词"to"开头。例如："Objective：To become a..."。求职目标一定要简短、明确——不超过2～3行。像"你能给公司带来什么"这种信息最好显示出来，比如一段特定的经历和技能。招聘者更关注你能给公司带来什么，而不是你希望从他们那里得到什么。发简历申请职位之前记得检查你的"求职目标"，确保它适合你所申请的职位。

如果不确定你想要什么工作，那么你可以用一个令人印象深刻的"个人概要"来代替。一份"个人概要"就是用一两句话来概括你所掌握的经验和技能。它突出强调使你成为一个优秀的求职者的原因，还有你与其他求职者的不同或者你超过他们的地方。书写时，要注意调整你的"个人概要"，凸显与你所申请工作的要求最切合的经验或者技能。

（3）任职资格（qualifications）

任职资格就是对自己的整体概括，根据求职意向用一两句概括自己最大的优势，最好有数字和细节证明。如应聘"电子工程师"的任职资格可以这样写:"累计4个多月电子公司实习经历;名牌大学电子工程本科教育背景,扎实的数学功底,全面的专业知识;熟练掌握 Word、Excel、PowerPoint 等 Office 系列办公软件;通过大学英语四、六级考试,能与他人较流利地进行英语对话。"

(4) 教育背景(education)

教育背景是简历中的一个重要信息,一般按照时间逆序的写法来写,主要是个人从大学阶段到毕业前所获得的学历,时间上需要衔接。学历从最高开始写,比如先写博士,后写硕士,再就是本科。高中阶段、初中阶段经历一般不写,但如果有获得特别的奖励或者与众不同的经历的话,如全国数学奥林匹克竞赛一等奖或者高考状元等,也可以写上。

在简历中,教育背景一般包括时间段、学校、学院或专业、学历等,也可写明自己的研究方向、导师、主修课程、辅修课程、研究项目、成绩排名、活动等。

(5) 工作经历(employment history/work experience/professional experience)或实践经历

这部分的书写要尽量突出你之前的工作经历与应聘岗位的符合程度,应届毕业生要突出自己的学习态度和各种可迁移的能力。平均学分成绩虽然不能说明全部问题,但是能证明你对学业是否认真,用人单位会看重。应届毕业生大多无具体的工作经验,用人单位会看重公司实习、担任学生干部、参与志愿服务、参与团队科研等各类社会实践活动经历。

每一段工作经历参考三段式描述,可按 Responsibility(承担工作职业和工作内容)、Achievement(取得的成绩)和 Learning(学习和收获什么)的模式来写。可将突出的工作经历或实习经验写在最前面。只写主要的工作经历,3~4 项就可以了,没必要把所参加的所有实践、项目等全部写出来,只需要描述与自己现在应聘职位要求所相关的经验、经历就可以了。工作经历的书写方法有两种:

1) 时序式(chronological style):

时序式即按逆时间顺序写。

2) 职务式(functional style):

职务式即按所担职务的不同分类写。

(6) 技能(skills)

这部分除了专业技能外,根据所谋职位需求,也可以突出其他个人技能。例如计算机技能、外语水平技能。用人单位很可能更看中求职者某个点而不是全部,比如你做过 Oracle 下的 Java 开发,而他们正好手头有这样的项目,你就会因为这一条被留下。所以,应该在简历里面将最拿手的、最切合求职岗位的写在前面,让对方知道你的强项。计算机技能和外语水平可以用获得的证书来作为有力的证明材料。

(7) 出版物(publications)

出版物分为著作(books)、论文(paper)和翻译(translation)三类,一般按出版时间的逆序或者顺序逐一排列。有时候也可以根据情况,按影响力大小进行排序。

(8) 专长与成就(specializations and accomplishments)

专长是专业范围内最突出最擅长的技能。填写专长时,应重点强调一个或两个方面的专业特长,一般不宜超过两项。填写成就时,一要实事求是,二要具体、定量。如获得什么奖励,参加过什么科研项目,做出哪些贡献,获得哪些发明创造方面的专利等,都可罗列于此。

（9）推荐人（references）

简历，特别是中文简历中一般需要推荐人，以证明自己在简历中介绍的情况是真实可信的，自己的品行和能力可以接受查询，某些人士可以对自己的情况予以介绍，提供证明，做出推荐等。这部分有三点应注意：一是要获得推荐人的允许和承诺；二是要附上他们现在的的通信地址、邮政编码、电话号码；三是要将该简历的复印件给他们各送一份，以便他们对简历所述有全面了解，能有的放矢回答询问。

3. Key Points for Resume Writing（简历写作要点）

1）语言简单明了，用概括性的语言说明作者申请入学或求职所具有的能力。

2）语句中不用第三人称 he, she（显得生硬）和第一人称 I（显得自命不凡），而较多使用的是动词性短语、名词性短语和形容词短语。

动词性短语　　e. g.：Learned how to operate a retail store.

名词性短语　　e. g.：Ability to organize, coordinate and supervise staff

形容词短语　　e. g.：Responsible for overseas sales of Textile Department.

3）简历不宜写得太长，篇幅以三页之内为宜，不写写作日期，通常选用标准 A4 纸。

4）资料必须是客观真实的，不要吹牛作假。

5）内容不要密密麻麻地堆在一起，项目与项目之间应有一定的空位相隔。

6）切忌不要写对求学或求职无用的东西。

4. Sample Resumes（个人简历例文）

Sample 1　For a college/university student without work experience（没有工作经验的大学毕业生）

RESUME

RICHARD ANDERSON

1234, West 67 Street,

Carlisle, MA 01741,

(123)-456 7890

E-mail: richard@163.com

OBJECTIVE: Electrical engineer/A position in research and development with an electrical engineering background. //An Electrical engineer in Technological Department with opportunities for advancement. //An Entry-level position（初级职位）in electrical engineering. Long range goal/Eventual goal is to become a project principal（项目负责人）in Technological Department. //An entry-level position responsible for computer programming. //An entry-level position in sales with eventual goal of manager of marketing department.

QUALIFATIONS: University major in/University education in Electrical Engineering involving special training in installation, operation and maintenance of electrical devices.

EDUCATION:

20XX–20XX Bachelor Degree/Master Degree/Ph. D. in Electrical Engineering & Automation, Department of Electrical Engineering, XX（大学名称）University.

Key major subjects taken included/Main courses included:

Fundamental of Electric Circuit Analysis（电路分析基础），Technology of Analogue Electronics（模拟电子技术），Technology of Digital Electronics（数字电子技术），Principle and Application of Microcomputer（微机原理及应用），Theory of Automatic Control（自动控制理论），English for Electrical Engineering（电气工程专业英语），Principle and Application of Single-Chip Microcomputer（单片机原理及应用），Technology of Computer Control（计算机控制技术），Computer Network（计算机网络），CAD of Electrical Circuit（电子线路 CAD），Technology of Power Electronics（电力电子技术），Fundamental of Electric Drive（电力拖动基础），Computer Simulation（计算机仿真），Technology of Engineer Inspection（工程检测技术），Electric Power Supply for Industrial Plants（工厂供电），Programmable Logical Controller System（可编程控制系统），Electric Measurement（电工测量），Theory of Electric Motors（电机学），AC Motor Regulating System（交流调速系统），Digital Signal Processing（数字信号处理）.

⋮
⋮

* AWARDS/HONORS：

XX University Scholarship	20XX – 20XX
Excellent Student	20XX
First-prize winner in Shanghai College Students Computer Contest	20XX

* SKILLS/ENGLISH SKILLS：

- Good/Excellent/Fluent written & spoken/oral English. Proficiency in oral and written English. Obtain CET – 4/6 certificate in 20XX/Possession of CET – 4/6 certificate. //Primary English. Able to have a dialogue in daily English and read English documents about major. Scored 607 on TOEFL in August, 20XX.
- Excellent/Strong/Good PC skills. Obtain intermediate certificate of computer application. Familiar with common software of office automation and database. Able to/Capable of C and PLC programming.

* EXTRACURRICULAR ACTIVITIES：

Vice-chairman, Student Union of Shanghai XX University, 20XX – 20XX.
Captain, University Football Team, Contributed to the winning of Shanghai College Students Football Contest, 20XX.

PERSONAL DATA：Birth Data：Sept. 18, 1994
Sex：Male/Female
Height：1.78 m
Weight：65kg
Health：Good/Excellent
Marital Status：Single/Married
Hobbies：Calligraphy, Badminton, Football

* **REFERENCES**: Available upon request.
REFERENCES WILL BE FURENISHED UPON REQUEST
注："/"和"//"分别表示可以选择的词汇和句子。带 * 的内容为可选项。

Sample 2　For college/university students with work experience（有工作经验的大学生）

RESUME

NAME　（Ms./Miss）

P. O. Box 88563	Born: June 12, 1996
Shanghai 200034, China	1.72M, 61kg, Married
Tel.：(021) 56783621 (o)　(021) 66883960 (h)	Excellent health

OBJECTIVE: Electrical engineer/Position as a design engineer in Electrical Engineering Department. //An electrical engineer position in Foreign Engineering Department with opportunities for advancement position in electrical engineering. Long range goal/Eventual goal is to become an administrator in Engineering Department.

QUALIFATIONS: University major in Electrical Engineering and/combined with/coupled with five years practical experience in installing, operating and maintaining of electrical installations. //Three years of successful job experience ranging from sales responsibilities to management of marketing department.

EXPERIENCE:

　　2006 - Present Assistant Engineer, x x x（单位名称）, Shanghai, 200012. Responsible for developing new type of electric motor and reduced cost by 25 percent through new design.

EDUCATION:

　　Bachelor of Science in Electrical Engineering, concentration of electrical machinery (2006), Shanghai XX University, Shanghai, China.

　　Note: Ranked second among the graduates of the academic year 2005 - 2006.

ENGLISH SKILLS:

　　Fluent in English.

COMPUTER SKILLS:

Languages: BASIC, FORTRAN, C/C + + , HTML
Packages: Matlab, AutoCAD, MS-Word, Excel
Systems: DOS, MS-Windows, UNIX, Linux

　　* **PUBLICATIONS**:

　　Over 20 journal and conference publications, numerous presentations. Some selected publications are:

1) Author (s). title of the article [J], Journal name, year, vol. #, no. #, pp. page range.
2) Author (s). title of the article [C], Conference name, year, conference place, pp. page range.
　　⋮

* **REFERENCES**: Available upon request.

PART III　Techniques of EST C-E Translation
（科技英语汉译英翻译技巧）

Unit 1　Techniques of Chinese Words Translation I（汉语词汇翻译方法 I）

随着科技的不断进步，与各国的贸易往来日益频繁，越来越多的公司和科研单位需要将国内的科技资料从汉语翻译成英语，以增强我国科学技术的推广和交流。而在将汉语翻译成英语的过程中，单句的翻译是基础。只要每个单句都写准确了，即使其他方面不够完美，也能够与外国读者沟通信息。因此，本书还以词汇翻译为基础，阐述了汉译英单句翻译的基本方法和技巧。

1. 冠词（Articles）

英语中没有汉语的量词，汉译英时量词经常省略。例如，三条鱼译为"three fish"；一轮满月译为"a full moon"。而汉语中没有冠词，所以，在汉译英时，中国学生最容易犯的错误就是漏用冠词和乱用冠词。

（1）一般应加冠词的情况

1）在单数可数名词前一般要加冠词，泛指时多用不定冠词。

Example 1：发射机通常是由几部分构成的。

译文：*A* transmitter commonly consists of several parts.

Example 2：这是一个 8V 的电池。

译文：This is *an* 8V battery.

a 和 an 的应用完全取决于不定冠词后面紧跟的第一个音素而非第一个字母，若是元音，则一定要用 an。

2）已提及的、特指的或带有后置修饰语的名词前加定冠词。

Example 1：当电流流过导线时，它会遇到一些阻力。这种阻力就称为电阻。

译文：When an electric current flows through a wire, it meets some opposition. *The* opposition is referred to as resistance.

Example 2：对系统的控制是一种跨学科的科目。

译文：*The* control of systems is an interdisciplinary subject.

（2）一般不加冠词的情况

1）泛指的物质名词或不可数名词前不加冠词，复数名词前不加冠词。

Example 1：电广泛应用在工农业中。

译文：*Electricity* is widely used in *industry* and *agriculture*.

Example 2：电能可由电动机转换成机械能。

译文：*Electrical energy* can be changed by *electric motors* into *mechanical energy*.

2）论文的标题、书籍名称等的冠词可以省去。

Example：《计算机入门》

译文：Introduction to Computers

3）专有名词一般不加冠词。

一般用单个词表示的国家名称或一个地点名称加"大学"构成的专有名称前不加冠词；由三个或三个以上的普通单词构成的单位或国家名称前要加定冠词。

Example 1：计算机科学系

译文：The Department of Computer Science

Example 2：中华人民共和国

译文：The People's Republic of China

Example 3：麻省理工学院

译文：The Massachussetts Institute of Technology

4）图题中一般省去冠词。

Example：图 2-5　安培表电阻对电路中电流的影响

译文：Figure 2-5　*Effect* of *ammeter* resistance on *current* in circuit

5）某些可数名词单数形式在泛指时可以省去冠词。

Example：欧姆首先发现了电流、电压、电阻之间的关系。

译文：Ohm first discovered the relationship between *current*, *voltage*, and *resistance*.

6）表示独一无二的人之前不加冠词，在人名所有格之前不加冠词

Example 1：欧姆定律；法拉第定律

译文：Ohm's law; Faraday's law

如果人名直接修饰普通名词，则一般在它之前加定冠词。

Example 2：戴维南等效电路

译文：*The Thevenin* equivalent circuit

(3) 特殊情况

1）在表示某个参数的单位的词前往往用定冠词。

Example 1：电位差的单位是伏特。

译文：The unit of potential difference is *the volt*.

Example 2：电容的单位是法拉。

译文：The unit of capacitance is *the farad*.

2）以下表达中必须用定冠词。

Any/None/Neither/Either/All/Most/One/Each/The rest + of the + 名词

Example 1：这两个齿轮没有一个啮合的。

译文：*Neither of the* two gears is engaged.

Example 2：惰性气体中无论哪一种都不会和其他物质化和而形成化合物。

译文：*None of the* inert gases will combine with other substances to form compounds.

3）几个名词并列时可以共用第一个名词前的冠词。

Example：电容取决于任何两个导体之间的尺寸、形状和它们之间的间隔距离。

译文：Capacitance depends on *the* size, shape, and separation between any two conductors.

4) 当表示"比较一下""计算一下""了解一下""做描述""做讨论"等时,在抽象名词前一般使用不定冠词。

Example 1: 先决条件是对电路的基本内容要有一个好的了解。

译文: The prerequisite is *a good knowledge* of electric circuit fundamentals.

Example 2: 本书的范围不允许对所有这些数学方法做详细的讨论。

译文: The scope of this book does not permit *a dtailed discussion* of all of these mathematical devices.

Example 3: 对该电路做定量的分析是相当复杂的。

译文: *A quantitative analysis* of this circuit is rather involved.

(4) 冠词的特殊位置

1) 定冠词的特殊位置。

all + the + 复数名词

both + the + 复数名词

Example 1: 我们实验室里的所有仪器都是国产的。

译文: *All the instruments* in our laboratory are home-made.

Example 2: 这两个参数在物理学中都很重要。

译文: *Both the parameters* are of great importance in physics.

2) 不定冠词的特殊位置。

too/so/as...as/how + 形容词 + a(an) + 单数名词

Example 1: 在实际应用中,电阻器的功率额定值这一特性往往与其阻值是同样重要的。

译文: In practical application, the power rating of a resistor is often *as important a characteristic as* its resistance value.

Example 2: 频率太低实际上会引起寄生频率成分。

译文: *Too* low *a* frequency may actually introduce spurious frequency components.

Example 3: 这台仪器能灵敏到测出压力的微弱变化。

译文: This is *so* sensitive *an* instrument that it can measure a slight change in pressure.

2. 数词(Numerals)

科技文献中通常有不少数词,数词的准确翻译直接影响到英文全句含义的正确性。通常情况下,大于零的小数值用文字表示,而较大和很小的数值用阿拉伯数字表示,不需要翻译。数词的翻译以简洁明了为原则,切忌译文烦琐,无法使人一目了然。数字的表达应遵循易读、易写、前后一致的原则,对于有统计意义的、与单位符号及数字符号连用的数字,一般都采用阿拉伯数字形式。关于数字的英文翻译一般规则如下:

1) 小于10的数字通常需全拼,但与单位缩写连用时,则应该用数字表示。例如: five sections, two-dimensional, one equation, tenfold, third 等需要全拼。但 11 equations, 13th, 11-fold, 5 ft, 5 m, 20/20 vision, a 7 km course, 250℃, latitude 60°13′14″N, 65±3Ma 等用数字表示。

2) 位于句首的数字或单词"number"需用全拼形式。例如:

Example 1: 已经将16种鲸鱼列入世界濒危动物的清单上。

译文: *Sixteen* species of whale have been added to the list of the world's endangered animals.

PART III　Techniques of EST C-E Translation（科技英语汉译英翻译技巧）　| 153

Example 2：数字 6 不应包含在总数中，数字 5 是序列中的最后一个。

译文：*Number* 6 should not be included in the total; *number* 5 was the last in the series. （not No. 6, or no. 5）

又比如：

Twenty-five of the three hundred samples were contaminated. （或写成 Of the 300 samples collected, 25 were contaminated）

译文：300 个样本中有 25 个受到了污染。

Sixty-eight percent of the reactor heat is...

译文：反应堆放出的热中 68% 是……

如果句首的数字超过 100，为了表达清晰，最好改写。例如：

Three hundred and sixty-seven participants took part in the study. 可改写为：

A total of 367 participants took part in the study.

译文：共计 367 人参与了研究。

3）数字形式与全拼形式不可混用。例如：

Nine out of *ten* samples; Groups of *8*, *52*, and *256* particles 等。

4）两个或多个具有修饰关系的数字连用时，通常需要全拼出其中的一个数字，以免混淆。例如：

300 *six-inch* core samples; ten *43 cent* stamps; one hundred twenty *1g* samples 等。

5）分数的译法。分数表示一般方式为分子（基数词）/分母（序数词）。例如：五分之三（three fifths）；十分之七（seven tenths）。

Example 1：月球的体积约为地球的 1/4。

译文：The moon is about *one-quarter* the size of the earth.

Example 2：月球的质量约为地球的 1/80。

译文：The mass of the moon is *1/80* that of the earth.

而"零点几""零点零几"等表示法，则使用分子为 a few/several，分母为 tenths/hundredths 等分数形式来表示。

Example 1：该电阻上的电压为零点几伏。

译文：The voltage across the resistor is a few (several) tenths of a volt.

Example 2：这只是汽化热的千分之几（即零点零零几）。

译文：This is only a few thousandths of the heat of vaporization.

此外，用英文表示很小的数值，一般分子用"基数词 + parts（基数为 1 时用单数形式）"来表示，分母用 per 或 in a + 阿拉伯数字。

Examples：百万分之三（three parts per (in a) million）；千分之七（seven parts per (in a) thousand）。

6）倍数表示法。

① 倍数的增长。

a. A 的大小（长度、质量……）是 B 的 N 倍。

下面有几种常用表达方式：

$$\begin{cases} \text{A is } n \text{ times as large (long, heavy...) as B} \\ \text{A is } n \text{ times larger (longer, heavier...) than B} \\ \text{A is larger (longer, heavier...) than B by } n \text{ times} \end{cases}$$

Example 1：铁的重量几乎是铝的三倍。

译文：Iron is almost three times *as* heavy *as* aluminum.

Example 2：氧原子的质量是氢原子的16倍。

译文：The oxygen atom is *16 times* heavier than the hydrogen atom.

Example 3：太阳的大小是地球的 3.3×10^5 倍。

译文：The sun is *330,000 times* as large as the earth.

也可以采用如下形式：

$$\begin{cases} n \\ n \text{ times} \\ x\% \end{cases} + \begin{cases} \text{the + 名词} \\ \text{that + 后置定语（常为 of 短语）} \\ \text{what 从句} \end{cases}$$

这一句型可看成在两部分之间省去了"as large/great,...as"部分，数词在句子中做前置修饰语。

Example 4：这个电压是加给放大器的信号的80倍。

译文：This voltage is *80 times* the signal applied to the amplifier.

Example 5：在发达国家，垃圾以大约为国民经济增长率的2.5～3倍的增长速度激增。

译文：Garbage is rapidly increasing in quantity at about *2.5 to 3 times* the increasing rate of national economy in developed countries.

Example 6：铝的电阻率为铜的1.6倍。

译文：The resistivity of aluminum is *1.6 times* that of copper.

Example 7：在这种情况下，患肺癌的危险性下降到吸烟者的30%～50%。

译文：In this case the risk of lung cancer falls to *30 to 50 percent* that of smokers.

b. "增加到 n 倍"或"增加了 $n-1$ 倍（净增加的部分）"的译法。在将汉语表示"净增加"的倍数译成英语时，要在汉语的倍数上再加一倍。

以 increase 为例，下面有几种常用表达方式：

$$\begin{cases} \text{increase (raise) } n \text{ times} \\ \text{increase by } n \text{ times} \\ \text{increase } n \text{ fold} \\ \text{increase by a factor of } n \end{cases}$$

Example 1：这根导线比那根长4倍。

译文：The wire is *five times* longer than that one.

Example 2：水银比水重约13倍。

译文：Mercury weights more than water by about *14 times*.

Example 3：钢产量比2004年增长了三倍。

译文：The output of steel has been increased *four times* as against 2004.

Example 4：若X增加到两倍，则Y增加到四倍。

译文：If X is doubled (increased two times), Y is increased *by a factor of 4*.

PART III　Techniques of EST C-E Translation（科技英语汉译英翻译技巧）　155

Example 5：这台机器改善了劳动条件，并使工效提高了三倍。
译文：The machine improves the working conditions and raises efficiency *four times*.
Example 6：在这种情况下，脉冲间隔及脉冲组长度均增加了七倍。
译文：In this case the pulse spacing and the group length are increased *by eight times*.
② 倍数的减少。汉语中不习惯讲"减少了若干倍"，而说"减少了几分之几"。例如"减少了 2/3"或"减少到 1/3"，译成英文就是"reduce 3 times"。除了 reduce 以外，还有若干表示"减少"的同义词、近义词，如 decrease、shorten、drop、cut down 等。以 reduce 为例，常用的表达方式有：

$$\begin{cases} \text{reduce by } n \text{ times} \\ \text{reduce } n \text{ times} \\ \text{reduce to } n \text{ times} \\ \text{reduce by a factor of } n \\ n \text{ fold reduction} \\ n \text{ times less (lower, lighter, weaker, shorter...) than} \end{cases}$$

Example 1：生产成本减少了 3/4（减少到原来的 1/4）。
译文：The production cost has reduced *four times*.
Example 2：这一方案的优点在于节约人工 4/5（节约到原来的 1/5）。
译文：The advantage of the present scheme lies in a *five fold reduction* in manpower.
Example 3：测试结果表明，这可将放射性降低到原来的 1/10 – 1/100。
译文：Tests indicate this might reduce radioactivity by *a factor between 10 and 100*.
Example 4：比起老式打字机，它主要的优点是质量减轻了 3/4。
译文：The principal advantage over the old-fashioned typewriter is a *four-fold reduction* in weight.

练习：

将下列句子翻译成英语。
（1）磁铁具有一个 S 极和一个 N 极。
（2）设计控制系统在很大程度上取决于复变量理论的应用。
（3）交流电路中的有用功率还取决于该电路中的电流和电压。
（4）若读者想要对数据通信有所了解，就必须对电传输特性有一个一般的了解。
（5）现在人们越来越认识到这一方法是很有价值的。
（6）这两台设备的质量都很好。
（7）必须确定移动这个物体需要多大的力。
（8）在电压相同的情况下，导线的电阻越大，流过的电流就越小。
（9）这根导线的电阻为零点零几欧姆。
（10）该数值约比理想值大 3.5 倍。
（11）不发达地区的总人口为 45 亿，占世界人口的 79%。
（12）来自电厂、汽车和工厂的二氧化硫的量是 25 年前的四倍。
（13）太阳中心的温度高达 1000 万℃。
（14）原子内第三层和最外层之间的一些电子层，含有多达 32 个电子。
（15）电压升高到 20 倍，电流强度就会降低到 1/10。

Unit 2　Techniques of Chinese Words Translation II
（汉语词汇翻译方法 II）

1. 介词

汉译英时正确使用介词，可以使英文语句更加精炼。以下是翻译中常用介词的一些特殊使用方法。

（1）of

1）of + 抽象名词 = 该抽象名词的形容词，但语气更强。

Example 1：这节所讲内容很重要。

译文：What is described in this section is *of great importance*.

Example 2：工程师们会发现，这本书作为一本有关基本问题的参考书是很有价值的。

译文：Engineers may find the book *of value* as a reference on basic problems.

2）of 可表示"在……之中"，既可用作最高级的比较范围，也可用在一般的句子中，但其后只能接可数名词。

Example 3：在实验室里的所有计算机中，这台的性能最好。

译文：*Of all the computers* in this laboratory, this one works best.

Example 4：在五种简单机器中，经常使用到滑轮。

译文：*Of the five simple machines*, the pulley is frequently used.

3）of 可表示前后两者处于同位关系。

Example 5：那个较轻的机器零件的质量为7kg。

译文：The lighter machine part has a mass *of 7 kg*.

4）of 后面的词是前面的词的逻辑主语。

Example 6：该曲线图表示了机油黏度随温度变化的情况。

译文：The graph shows the variation *of engine oil viscosity* with the changed temperature.

5）of 后面的词是前面的词的逻辑宾语。

Example 7：一个力可以分解成 x 分量和 y 分量。

译文：*The resolution of a force* into x-and y-components is possible.

（2）with

1）with + 抽象名词 = 该名词对应的副词，但语气更强。

Example 1：这些实验应仔细地做。

译文：These experiments should be done *with care*.

2）with + vary/change/increase/decrease/… 表示"随着……"。

Example 2：半导体的导电率随温度变化。

译文：The conductivity of a semiconductor varies *with temperature*.

3）with + 复合宾语结构可以视为无动词从句或非限定从句，这种结构可做状语表示原因、条件、让步等。

Example 3：由于存在摩擦，一部分功率作为热损耗掉了。
译文：*With friction present*, a part of power has been lost as heat.
Example 4：当温度不变时，空气的密度同压力成正比。
译文：The density of the air varies directly as pressure, *with temperature constant*.
4）with 用在句首表示"对于，有了"等意思。
Example 5：对于交流电来说，情况就不同了。
译文：*With the alternating current*, things are different.
Example 6：在作用力不变的情况下，加速度和质量成反比。
译文：*With a constant force*, the acceleration is inversely proportional to the mass.
（3）by
1）by 可表示除了时间和距离外的任何参量的数值（也可以把它省去）。
Example 1：在这种情况下，v 和 i 的相位相差 90°。
译文：In this case v and i differ in phrase by 90°.
Example 2：尽管汽油价格降幅过半，但到上周为止消费依然走低。
译文：Despite a drop *by more than half* in the price of gasoline, consumption until last week remained low.
2）by 可表示"按照，根据"，主要用在推导中。
Example 3：通过考察该设备的性能，我们就能了解它的特点。
译文：*By an examination of the performance of the device*, we can understand its features.
（4）for
在 method, technique, equation, algorithm, condition, requirement 等词后面多用 for。例如：
Example 1：天线的设计方法有几种。
译文：There are several *methods for* antenna design.
Example 2：通过理论分析，提出了解决此类问题的智能算法。
译文：By a theoretical analysis, the intelligent *algorithms for* solving this kind of problem are put forward.
（5）on 和 upon
on/upon + 动名词或表示动作的词时表示"一……就……"或"在……之后"。
Example：物质一经挤压，其体积就会变小。
译文：*On being compressed*, the volume of the substance will be reduced.
（6）in
1）in 后跟表示单位的名词复数（hertz 除外）时，意为"用"。
Example 1：频率是以赫兹为单位来度量的。
译文：Frequency is measured *in hertz*.
2）in 后跟表示方向的词时，意为"朝"。
Example 2：无线电波朝四面八方传播。
译文：Radio wave travels *in all directions*.
3）在名词 reduction decrease, increase, change, drop, rise, fall 等词后多用 in。

Example 3：这些设备的体积各不相同。

译文：These devices differ greatly *in size*.

Example 4：我们能够测出压力的微弱变化。

译文：We can measure the slight change *in pressure*.

4) in 后面跟动名词或表示动作的名词，意为"在……时候；在……过程中，在……方面"。

Example 5：在讨论微分方程的时候，我们将把注意力限于一次方程上。

译文：*In our discussion of differential equations*, we shall restrict our attention to equations of the first degree.

（7）over

1) over 表示"通过……，越过……"时，其后常跟表示距离的名词，其前常为表示移动的动词。

Example 1：毫无疑问，这只鸟一次能飞很长的距离。

译文：The bird can surely fly *over* a long distance without stop.

2) over 表示"与……相比"时，通常在 advantage 等词后。

Example 2：新方法与以前的那些方法相比有了很大的简化。

译文：The new method has an enormous simplification *over* the previous ones.

3) over 引出个别具有及物动词含义的名词的逻辑宾语。

Example 3：在冰上开车时，他尽力控制住了汽车。

译文：He managed to *keep control over* his car on the ice.

2. 副词

（1）副词做后置定语

above, below, here, there, around, nearby, up, down, away, apart 等通常做后置定语。

Example 1：下面的表格列出了物质的电阻率。

译文：The table *bellow* lists resistivities of some substances.

Example 2：由于机器发出的噪声两个人相距 5m 并不能互相听清对方的讲话。

译文：Two people *five meters apart* are unable to hear each other clearly due to the noise made by machine.

时间副词 now, then, today, afterward 等通常做后置定语。

Example 3：现在的问题是要确定电流的大小。

译文：The problem *now* is to determine the magnitude of the current.

（2）副词做状语

翻译中将副词做状语时，要注意正确放置副词的位置。

有些副词可以放在句首。

Example 1：习惯上流向器件的电流被指定为正。

译文：*Conventionally* current flowing toward a device is designed as positive.

不及物动词与介词连用的情况下，副词往往放在介词与动词之间。

Example 2：电动势随着使用仅仅稍有下降。

译文：The EMF decreases *only slightly* with use.

副词放置在主动语态中主动词之前。

Example 3：这个图清楚地表明了测试点。

译文：The figure *clearly* shows the test points.

副词放在宾语之后。

Example 4：我们可以同时解出这些方程。

译文：We can solve the equations *simutaneously*.

副词放在及物动词之后。

Example 5：我们假定在电路中只存在电流 I。

译文：We assume *only* that a current I is present in the circuit.

系表结构中，副词一般放在 is, are 之后。有 will, can, may 则放在 be 之前。

Example 6：反向电压增益通常是可以忽略不计的。

译文：The reverse voltage gain is *usually* negligible.

Example 7：结构图的尺寸必须准确。

译文：Structure drawings must *dimensionally* be correct.

副词在被动句中可以直接放置在 be 之后，又可以放置在 will 和情态动词之后，be 之前。或放在所有动词之后，即行为动词谓语之后。

Example 8：该噪声电压可以大大地加以衰减。

译文：The noise voltage can be *greatly* attenuated.

Example 9：这个参数也可以从伏安特性曲线获得。

译文：This parameter can *also* be obtained from the *I-V* characteristic curve.

Example 10：两组继电器可同时被打开或闭合。

译文：Two relays can be opened or closed simultaneously.

练习：

将下列句子翻译成英语。

(1) 这个参数可以精确地加以测量。

(2) 有了雷达，我们就可以看见远处的物体。

(3) 通过分析这个模型，人们能了解原子的结构。

(4) 直流只朝一个方向流动。

(5) 电容器的电容取决于平板的大小及其分开的距离。

(6) 附近的建筑物大部分是现代化的。

(7) 电子计算机的主要特点是计算准确而快速。

(8) 电子开关并不能即刻执行"或"和"与"逻辑操作。

(9) 最小的门电路输入电压将可靠地被看作逻辑1。

(10) 他们将用电子邮件将有关这篇论文的信息发送给世界上所有重要的科学家。

(11) 爆炸后的强大气浪把方圆几英里之内的房屋全部夷平。

(12) 通过该系统的应用，实现了生产过程的自动化和标准化。

(13) 采用数字信号技术已将火星图片成功传输到地球。

(14) 现在已发明了一种自动点煤气炉的装置。

(15) 电动式燃油泵比机械式燃油泵更有优势。

Unit 3　Techniques of Chinese Words Translation III
（汉语词汇翻译方法 III）

1. 动词

在科技文献翻译的过程中，动词的翻译应注意以下几点：

（1）get，turn，go，stay，appear，look，prove 等特殊连系动词的使用

Example 1：当电流流过导线时，该导线会发热。

译文：When an electric current flows through a wire, the wire will *get* hot.

Example 2：在这种情况下，输出保持高电位。

译文：In this case, the output *stays* high.

Example 3：远程学习使一些大学成为全国性的或国际性的大学。

译文：Tele-learning makes some universities *go* national or international.

（2）半助动词（remain，seem，appear，happen 等）与动词不定式合成谓语

Example 1：这些实验似乎表明世界上只有两种电荷。

译文：These experiments *seems* (*or appear*) *to* indicate that there are only two kinds of electric charge in the universe.

Example 2：另外，毫无疑问在呼吸中还有大量更多的疾病病因等着被发现。

译文：In addition, lots more markers of disease in the breath no doubt *remain to* be discovered.

Example 3：关于进化的事实哪一些肯定是真实的，而哪一些碰巧是真实的呢？

译文：Which facts about evolution to be true and which *happen to* be true?

Example 4：这些问题对人类的影响有待于研究。

译文：The effects of these problems on human beings *remain* to be studied.

（3）助动词（do，does，did）主要出现在比较状语从句和方式状语从句中

Example 5：这些无线电波的性能像光波一样。

译文：These radiowaves behave as light waves *do*.

（4）动词的省略

在科技文献中常常遇到一类动宾短语，像调整位置（setting）、编制程序（programming）、配置仪器（instrumentation）、加负载（loading）、定向（orientation）等，常译为相关的动名词或具有动作意义的名词。这时，从字面上看，像是省略了其中的述语动词。通常略去不译的这类汉语动词有：加、求、使、进行、实行、产生、引起、发生、成立、采用、发挥、确定、做出、制定、提高等。

Example 1：解决捕获、跟踪和识别所有目标的问题，要求发挥相控阵雷达的全部能力。

译文：The *problem* of acquiring, tracking and discriminating all the targets needs the full *capability* of the phased array radar.

Example 2：在宇宙飞行的伟大探险活动中，下一步是建立宇宙空间站。

译文：The next stage in the great adventure of space travel is a *space station*.

Example 3：面对专业化，不可避免地应采取相应的组织措施。

译文：The inevitable counterpart of specialization is *organization*.

Example 4：这是打破了纪录的增长数字。

译文：This is a *record* increase.

还有一些"动词+非名词宾语"结构中的动词可以省略，因为此类结构通常使用的动词是"进行""开展""加以""予以""禁止""感到""严加"等。英译时通常省略这些动词，而只采用其后的"非名词宾语"——另一动词。例如，将"进行研究"译成 to design等。

2. 形容词

（1）形容词做后置定语

形容词 present，else，what ever 通常要后置。

Example 1：这个电荷与存在的其他电荷相互作用。

译文：This charge interacts with other charges *present*.

Example 2：即使在低压下，仍存在着大量分子。

译文：Even at low pressure there are still large numbers of molecules *present*.

形容词 available，obtainable，achievable，responsible，possible，usable，inclusive 等放在被修饰的名词后面以加强语气。

Example 3：这些是所能获得的最小微粒。

译文：These are the smallest particles *obtainable*.

Example 4：这一措施是能获得极低直流功率的关键。

译文：This measure is the key for the extremely low DC power *available*.

形容词只能放在由 some，every，any，no 与 thing，body，one 组成的不定代词后面。

Example 5：现在计算机没什么神秘的了。

译文：Now there is nothing *mysterious* about computers.

Example 6：所有的数字设备都将数字化。

译文：Everything *electronic* will be done digitally.

（2）形容词短语做后置定语。

由 and 或 both... and 以及 or 或 either... or 连接的两个形容词可以做后置定语。

Example 1：本书对新老电路设计人员都是有帮助的。

译文：This book is a help to circuit designers *both new and old*.

Example 2：该放大器的效率为50%，这是理论上所能达到的最大值。

译文：The efficiency of the amplifier would be 50 percent, the maximum *theoretically possible*.

Example 3：无穷大是大于任何数的一个量。

译文：Infinity is a quantity *greater than any number*.

（3）形容词短语做状语

Example 1：这种设备由于结构简单、价格低廉，所以需求量很大。

译文：*Simple instructure and low in price*, this device is in great demand.

Example 2：所有的电路不论大小，都含有同种元件。

译文：*Large or small*, all the circuits will contain the same kinds of components.

Example 3：如果铁块不受潮，则不易生锈。
译文：*Free from the attack of moisture*, a piece of iron will not rust very fast.
练习：
将下列句子翻译成英语。
（1）每个物体不论大小都具有引力。
（2）中子既不带正电，也不带负电。
（3）电流等于电源电动势除以电路的总电阻，既包括外电阻也包括内电阻。
（4）把弹簧拉得越长，拉伸所需要的力就越大。
（5）在导线内能自由运动的电子在形成电流方面起了极为重要的作用。
（6）电子技术对计算技术的影响极其深远。
（7）如果铁块不受潮，则不易生锈。
（8）与大家的观念相反，力并不仅仅是靠直接接触来传递的。
（9）必须注意，电流与电阻成反比。
（10）该导体正平行于磁场运动。
（11）看来海洋生物数量巨大，品种繁多。
（12）这种情况下不会发生任何转动。
（13）负责部门将采取具体措施来制止污染。
（14）在测量远短于1s的时间间隔时，就采用十进制。
（15）设计任务起始于一种需求，无论这种需求是真实的还是想象的。

Unit 4　Techniques of Chinese Words Transaltion IV（汉语词汇翻译方法 IV）

1. 代词

（1）物主代词做定语

Example 1：一只理想的安培表具有非常低的等效电阻，以使它存在于任一电路中时，该电路的性质发生改变的程度尽可能地小。
译文：An ideal ammeter has a very low equivalent resistance so that *its* presence in any circuit alters the properties of the circuit as little as possible.
Example 2：有关收敛问题并不简单，对它们的研究构成了现代分析中重要的一章。
译文：The questions of convergence are not simple, and *their* study forms an important chapter in modern analysis.
Example 3：人们用肉眼看不到细菌。
译文：One cannot see bacteria with *one's* naked eyes.

（2）one 的用法

用 one 代替"有人，大家，人们"。
Example 1：所谓最有效的方法，人们一般是指最快的算法。

译文：By the "most efficient" algorithm *one* normally means the fastest.

Example 2：人们往往认为摩擦是不希望有的。

译文：*One* often thinks that friction is undesirable.

有时汉语中把修饰的定语放在名词的前面，这样就变成了一个句子，而在译成英语时，如果能够将这个句子拆开而意思不变，常常需要用 one 一类的词做 which 的先行词。例如：

Example 3：标量是只计大小的量。

译文：A scalar quantity is *one* which is completely *defined* by its magnitude alone.

（3）用 it, its, they, their 等词来代替句中的人或物。

Example 1：计算机在工作之前必须被告知要做什么。

译文：Before *it* can work, a computer must be told what to do.

Example 2：金的价格昂贵，因此不能广泛用作导体。

译文：The cost of gold is high. That is why *it* is not extensively used as conductors.

2. 名词

（1）表示大小尺寸等名词做后置定语

Example 1：这里 m 是不为零的任意数。

译文：Here *m* is any number *not zero*.

Example 2：他们安装了一台如茶壶大小的 250W 发电机。

译文：They installed a 250 W generator *the size of teapot*.

（2）单个名词做方式状语

Example 1：晶体管化的小型设备往往由电池供电。

译文：Small transistorized equipment is often *battery* powered.

Example 2：这台设备可用计算机来控制。

译文：This device can be *computer* controlled.

Example 3：需要对该设备进行空气冷却。

Air cooling the equipment is necessary.

（3）名词短语做同位语

Example 1：波能绕其通道上障碍物的边缘弯曲前进，这一特性称为绕射。

译文：Waves are able to bend around the edge of an obstacle in their path, *a property called diffraction*.

Example 2：电阻表是测量电阻的仪表，它被广泛应用于电气工程中。

译文：*An instrument for measuring electric resistance*, the ohmmeter is widely used in electrical engineering.

（4）名词与介词的搭配

Example 1：太阳离地球的距离是很远的。

译文：The *distance* of the sun *from* the earth is great.

Example 2：速度被定义为距离与时间之比。

译文：Speed is defined as the *ratio* of distance *to* time.

Example 3：这条曲线画出了该电路的电流随外加电压的变化情况。

译文：The curve shows the *variation* of the current in the circuit *with* the applied voltage.

(5) 特殊名词的复数形式

Example 1：制造商提供了各种各样的设计用以实施适当接口的集成电路。

译文：Manufacturers provide a large variety of *IC's* designed to effect proper interfacing.

翻译时需要注意缩略词的复数形式，再如：

Example 2：我们使得所有的 *R* 和 *C* 均相等。

译文：All the *R's* and *C's* are made equal.

Example 3：计算机能把 0 和 1 的同样的二进制排列翻译成数据或翻译成一条指令。

译文：The computer can interpret the same binary configuration of *0's* and *1's* as data or as an instruction.

Example 4：在这个数据单中给出的转换时间表示了输出波形的上升和下降时间。

译文：The transition time given in this data sheet represents the rise and fall times of the output waveform.

(6) 名词的增加

在汉语里，有时说出中心词就完全可以把意思表达清楚了，但英译时却必须补全该中心语的修饰语（定语）。办法往往是利用介词 of 译作后置定语，而在 of 前面则需添加表示度量等意义的名词。

Example 1：不同的物质具有不同的特性。

译文：Different *kinds of matter* have different properties.

Example 2：金属经过热处理后，强度更大，更加耐用。

译文：Going through *the process of heat treatment*, metals become much stronger and more durable.

(7) 名词的省略

科技英文中经常用到如"研究工作""解决办法""滞后现象""折中方案"等定语中心短语形式的名词词组。这类短语的定语部分，是由具有动作意义的抽象名词担任的，中心语是与该动作有关的。这类短语往往采用派生出来的行为动作的抽象名词译出其中的定语。其中一部分中心语经常略去不译。比如这些词可以不译：量、值、作用、现象、系统、方法、设备、结果、过程、研究、技术、形式等。

Example 1：根据水的蒸发现象，人们知道液体在一定条件下能变成气体。

译文：From the *evaporation* of water people know that liquids can turn into gases under certain conditions.

Example 2：电工技术为我们提供电取暖器、电照明设备、电吹风、电唱机和许多或许你家中就有的电气设备。

译文：Electrical technology gives us *electric heating* and *lighting*, hairdryers, record players, and many of the gadgets you probably have in your home.

名词在翻译的过程中省略还有一种情况就是为了使语句准确而精炼，有时可以省略形容词前面的名词。例如，汉语中像"价廉物美"这种形式的主谓短语（形容词担任谓语）用在句子中时，译成英语时往往省略了前面的名词，只译出其中的形容词就够了。

Example 1：标准的汽油发动机具有重量轻、容易制造的优点。

译文：A normal petrol engine has the advantages of being *light* and easily constructed.

PART III　Techniques of EST C-E Translation（科技英语汉译英翻译技巧）

Example 2：这部打字机真是物美价廉。

译文：This typewriter is indeed *cheap and fine*.

Example 3：一种新型飞机正越来越引起人们的注意——这种飞机体积不大，价钱便宜，无人驾驶。

译文：A new kind of aircraft—*small*, *cheap*, *pilotless*—is attracting increase attention.

（8）名词短语代替表示条件、原因、目的、时间等的状语从句。

Example 1：如果人们对微积分做进一步的研究，就会发现幂级数的其他许多用途。

译文：*A further study of calculus* shows many other uses of power series.

此句中名词短语"A further study of calculus"代替了条件状语从句"If one makes a further study of calculus"。

Example 2：如果Q值太大，就会损坏该仪器。

译文：*Too large a value of Q* would damage the device.

此句中名词短语"Too large a value of Q"代替了条件状语从句"If the value of Q were too large"。

Example 3：因为太空中没有空气，所以科学家们可以在那里制造出纯净的药来。

译文：*The absence of atmosphere in space* would enable scientists to make pure drugs there.

此句中名词短语"The absence of atmosphere in space"代替了原因状语从句"Since there is no atmosphere in space"。

Example 4：加上反馈后，输出变稳定了。

译文：*The addition of the feedback* makes the output stable.

此句中名词短语"The addition of the feedback"代替了时间状语从句"After the feedback is added"。

练习：

将下列句子翻译成英语。

（1）在人们学习某一系统前，必须定义和讨论一些重要的术语。

（2）立方厘米代表了各边长为1cm的立方体的体积。

（3）这个控制单元是受指令控制的。

（4）小汽车加速时车头会向上抬起，这是大家熟悉的一种效应。

（5）二极管的主要作用之一是在于从交流电源产生出直流电压，这一过程称为整流。

（6）本章讨论温度对晶体管的影响。

（7）这些特殊问题是把原子能用作能源时出现的。

（8）要注意所有的"g"在计算中都消掉了。

（9）控制信号能够在不同的时刻到达而不影响输出状态。

（10）应注意保证脉冲信号本身不出现不规则现象和中断现象。

（11）下面我们将讨论把白色的光分解成光谱的各种颜色。

（12）从水中吸走足够的热量可能使水变成冰。

（13）由于篇幅有限，我们只能简单讨论一下石英晶体技术。

（14）y 对于 x 的依从关系被称为函数。

（15）让我们把心脏比作一个泵来了解它是如何工作的。

Unit 5　Translation of Emphatic Sentences
（强调句的翻译）

为了突出句子中所表达的某部分内容，汉语与英语一样，都要采用强调的语气。但汉语和英语用来表达强调的方法有很大的差异。通常将汉语强调句翻译成英语句子的方法有以下几种。

1. 译成 "… + do/does/did + 动词原形 +…" 强调句

将助动词"do"置于动词谓语前面，即构成了英语中对谓语加以强调的强调句型。汉语谓语前面有"是""的确是""确实""确实是"之类的副词表示强调都可以用这种句型翻译。根据时态的要求，"do"可以演化出"does""did"两种形式，而其后的动词则用动词原形。

Example 1：他的确生产出一种便宜的玩具，但因其结构简单，不能变换成许多图形，因而卖不出去。

译文：He *did* produce a cheap toy, but as its design was simple and couldn't be changed into many figures, the product couldn't be sold.

Example 2：那些确实完成实验的人是知道脑电信号采集中的注意事项的。

译文：Those who *did* complete the experiment know what should be considered during the acquisition of EEG signals.

Example 3：这样，我们就证明了蜜蜂确实把蓝色认定为一种颜色。

译文：In this way we have proved that bees *do* really see blue as a color.

Example 4：一个半世纪以前，人们对于恐龙还一无所知，这事说起来真有些不可思议。

译文：It *does* seem strange to think that only a century and a half ago dinosaurs were quite unknown.

Example 5：如果正电荷确实在导线中运动的话，它们就从正端流向负端。

译文：If the positive charges *did* move in a wire, they would flow from the positive terminal to the negative one.

2. 译成 It + be + 谓语之外的成分 + that

（1）强调主语

Example 1：是他在有机化学中做出了如此伟大的发现。

译文：*It was he who* made such a great discovery in organic chemistry.

Example 2：是太阳提供了我们所使用的大部分化学能。

译文：*It is the sun that* gives most of chemical energy we use.

Example 3：使机车开动的正是燃烧燃料所产生的热。

译文：*It is the heat from burning fuel that* makes the locomotive run.

（2）强调宾语

Example 1：现今，人们在大多数机工车间里用以驱动机床的就是电动机。

译文：Nowadays, *it is the electric motor that* man uses to drive machines in most of the machine shops.

Example 2：我们用来隔音的<u>就是</u>纤维板。

译文：*It is fibreboard that* we employ to insulate sound.

Example 3：我们必须想各种办法来克服的<u>正是由摩擦引起的损失</u>。

译文：*It is the loses caused by friction* that we must try to overcome by various means.

（3）强调状语

Example 1：我们<u>是从太阳那里</u>得到了光和热。

译文：*It is from the sun* that we get light and heat.

Example 2：中国实验火箭达到这样的高度<u>还是第一次</u>。

译文：*It was the first time* that a Chinese experimental rocket had reached such a height.

Example 3：人们<u>正是在这里</u>首次开动计算设备，利用无线电观测结果昼夜不停地进行计算，预报了一切发射卫星的轨道。

译文：*It was here* that around-the-clock computational facilities were first maintained for the prediction, from radio observations, of the paths of all launched satellites.

Example 4：要靠<u>这些</u>模具才能使塑料制品形成各式各样的形状。

译文：*It is the moulds* which give plastic objects their shapes.

3. 译成"It is not until... that"强调句

Example 1：<u>直到1959年</u>，化学工作者才成功地获得了这种化合物。

译文：*It was not until 1959* that chemists succeeded in obtaining this compound.

Example 2：<u>直到1974年世界油价提高了300%的时候</u>，风机才重新被看作是一种替代能源。

译文：*Not until the 300 percent increase in world oil prices in 1974* were wind turbines again envisioned as alternative source of energy.

4. 译成部分倒装句

Example 1：我们<u>只有掌握了丰富的科学知识</u>，才能在现代社会中获得成功。

译文：*Only by obtaining a good knowledge of science*, can we live successfully in modern society.

Example 2：<u>只有在这种情况下</u>我们才在该电阻器的两端跨接一个电容器。

译文：*Only in such a case* is a capacitor connected across the resistor.

5. 采用形容词"very"来强调某个名词

Example 1：交流电<u>就</u>是使无线电成为可能的<u>那</u>种电流。

译文：The alternating current is the *very* current that makes radio possible.

Example 2：这样，当把仪表介入电路后，它就不会改变我们想要测量的<u>那个</u>参量。

译文：In this way, when the instrument is inserted it does not change the *very* thing we wish to measure.

练习：

将下列句子翻译成英语。

（1）也正是电子学为确确实实在创造奇迹的计算机赢得了"电脑"的称号。

（2）事实上，由于科技英语用于专业文章中，加上科学家和工程师们对专业的兴趣，因此，科技英语确实和普通英语不同。

(3) 正是当物体受热时，分子的平均速度提高了。

(4) 正是测量一方面把科学与数学联系起来；另一方面又把科学与商业和力学方面的实践联系起来。

(5) 我们必须想各种办法来克服的正是由摩擦引起的损失。

(6) 正是从这个意义上，数学有时被称为科学的语言。

(7) 人类就是利用电子学来设计自动控制装置的。

(8) 只是因为有了变压器，交流电压才能升高或降低而能量损耗很小。

(9) 就是在我们闭合电路的那一瞬间电流才开始流动的。

(10) 直到1925年才证实了电离层的存在。

Unit 6 Translation of Attributive Clauses
（定语从句的翻译）

科技英语为了准确描述客观现象、实验结果、生产流程或物质特性等，往往广泛使用各种从句，其中最常见的是定语从句。使用定语从句使句子结构变得复杂，形式冗长，因而承载的信息量也随之增加，便于描述客观事物的复杂性。但是，这也使英语的定语从句成了各种从句中最复杂、最难译的一种从句。

英语的定语从句是由关系代词或关系副词引导的从句，在句中做定语，一般是对某一名词或代词进行修饰和限制，被修饰的词被称为先行词，而定语从句位于先行词之后。在结构形式上有许多种，如直接由关系代词、关系副词引导的，由"介词+关系代词"引导的，由"名词+介词+关系代词"引导的，由"介词+which+名词"引导的，以及许多割裂式的定语从句。在汉译英过程中，如何清晰地表达原文的科技含义，并避免冗长的定语从句修饰则是我们在翻译中需要注意的。下面就来总结一下定语从句的翻译方法：

1) 关系代词在从句中做主语，宾语和定语以及关系副词在从句中做状语。

汉语中，如果同一主语在并列的两个简单句内先后两次出现，在后一句中这个名词习惯上要加以重复，而译成英语时，可以用一个which引导的定语从句来代替，从而避免该名词的重复。

Example 1：当实际用户单击按钮时，它会生成一个Windows消息，该控件对该消息进行处理，并将其转换成管理事件。

译文：When a real user clicks on a button, it generates a Windows message which is processed by the control and turned into a managed event.

还有其他一些关系代词和关系副词的使用，可以使英文语句更加紧凑，而避免词语的重复。

Example 2：我们必须懂得在这里使用倍频电路的理由。

译文：We must understand the reason *why the frequently multiplier circuit is used here.*

PART III　Techniques of EST C-E Translation（科技英语汉译英翻译技巧）

Example 3：目标就是其位置要加以确定的物体。
译文：A target is the object *whose position is to be determined*.
Example 4：能用来测量电流、电压、电阻的仪器叫作万用表。
译文：The instrument *that can be used to measure current, voltage and resistance* is called a multimeter.

2）关系代词在从句中做介词宾语，而"介词+which"在从句中做状语的句型。这种句型选择介词应从三个方面加以考虑：与名词的搭配要求；从句中动词、名词或形容词所需的搭配要求；以及整个句子所要表达的概念。

Example 1：做功的速率被称为功率。
译文：The rate *at which work is done* is referred to as power.
Example 2：电磁感应是产生几乎世界上所有电力的方法。
译文：Electromagnetic induction is the means *by which nearly all the world's electric power is produced*.
Example 3：动物呼吸空气，通过肺把其中的氧气吸收到血液里。
译文：Animals inhale air *from which the oxygen is absorbed*.
Example 4：在后面一章我们将看到不同的光源发射出不同类型光谱的理由。
译文：In a later chapter we will see the reason *for which different sources emit different types of spectrum*.
Example 5：构成水的两种元素是气体氧和氢。
译文：The two elements *of which water consists* are the gases oxygen and hydrogen.
Example 6：在电流流过的所有电阻中，电能被转换成了热能。
译文：In all resistances *through which a current flows*, electrical energy is converted into heat energy.

3）先行词为不定代词或被其他词修饰时。先行词为不定代词或被序数词、形容词最高级或形容词 only, no, very 等修饰时不能用 which，而只能用 that，不过在 something 后英美人也有用 which 的。

Example 1：这是我们所能采取的唯一措施。
译文：This is the only measure *that we can take*.
Example 2：人们只需要按一下按钮。
译文：All *that one need* do is push the button.
Example 3：计算机是人类所曾有的最有效的助手。
译文：Computers are the most efficient assistants *that man has ever had*.
Example 4：这样，接入的仪表并不会影响我们想要测量的参数。
译文：In this way the inserted meter will not affect the very thing *that we wish to measure*.

4）在科技文章中关系词可省去的两种主要场合：

第一种场合：关系代词在从句中做及物动词的宾语时（该动词不一定是谓语）可省略，这样可以使英语语句更加简练和紧凑。

Example 1：雷达是通过测量无线电回波到发射源所需的时间来检测目标距离的一种仪器。

译文：Radar is an instrument that measure the distance of a target by measuring the time *a radio echo takes to return to the source*.

Example 2：每个中央处理装置具有一套它懂得如何执行的极为基本的功能。

译文：Each CPU has a very elementary set of functions *it knows how to perform*.

Example 3：这个部件执行我们给它发出的指令。

译文：This component carries out the instructions *we give it*.

Example 4：为了满足这一条件，我们只需要把积分路径移到 σ 的右边。

译文：To satisfy this condition, all *we need do* is to shift the path of integration to the right of σ.

Example 5：一个物体产生的万有引力的大小，取决于它所含物质的多少。

译文：The amount of gravitational pull *a body produces* depends on the amount of material in it.

第二种场合：在 way，distance，direction，reason，time，number of times，amount 等后可以省去关系副词或"介词+which"，这时也可用关系副词 that 来引导从句。

Example 6：功等于力与物体运动距离之乘积。

译文：Work is the product of the force and the distance *a body moves*.

Example 7：阻尼振荡取决于开关闭合的时间。

译文：The damped oscillation depends on the time *the switch is closed*.

Example 8：这个图说明了电流随时间的变化情况。

译文：This diagram shows the way *the current changes with time*.

Example 9：在这种情况下成为电流反馈的理由在于流过 R 的电流大小确定了反馈的百分数。

译文：The reason *this is called current feedback* is that the amount of current flowing through *R* determines the percentage of feedback.

Example 10：回波只能从处于天线所指的方向上的物体那里反射回来。

译文：Reflections come back only from objects in the direction *the antenna is pointed*.

5）which 引导修饰整个主句的非限制性定语从句，这时的 which 等效于 this 或 that。汉译英时，这种"补缺which"似乎应用得更为广泛。这是因为在汉语的连贯叙述中，本来应该在两个分句内先后出现两次的名词，在后面的叙述中，为了语气的通顺连贯，避免文笔拖沓，根据汉语的习惯，这个名词常可省略，但汉译英时，则需要用 which 来补缺。which 引导修饰整个主句的非限制性定语从句通常有如下三种情况：

第一种情况：which 在从句中做主语。

Example 1：调谐放大器可以抑制远离谐振频率的信号，这往往是一大优点。

译文：A tuned amplifier rejects signals far from the resonant frequency, *which is often a considerable advantage*.

Example 2：可以把输入连接到其两个端点皆不接地的信号源上，这往往证明是方便的。

译文：The input may be connected to signal sources that have neither terminal grounded, *which often proves to be convenient*.

Example 3：能量的一部分被转化为热，（这种热）来帮助维持体温在37℃左右。

译文：Some of the energy is converted into heat, *which helps to maintain our body temperature*

PART III   Techniques of EST C-E Translation（科技英语汉译英翻译技巧）

at about 37℃.

Example 4：某些材料容易受潮，这会降低它们的绝缘性能。

译文：Some materials are liable to absorb moisture, *which will adversely affect their insulating properties.*

第二种情况：which 在从句中做介词宾语的定语。

Example 5：物理学、电子学、化学等的许多定律都是以代数式的形式表示的，在这种情况下容易地使用数字计算机。

译文：Many of the laws of physics, electronics, chemistry, etc. are expressed in the form of an algebraic formula, *in which case digital computers may be easily used.*

Example 6：这一传递过程一直延续到温度均衡时为止，那时能量便不可能进一步传递了。

译文：This transfer continues until a uniform temperature is reached, *at which point no further energy transfer is possible.*

Example 7：这种扩散电流持续到达到平衡为止，此时内部的势垒电位就建立起来了。

译文：This diffusion current continues until equilibrium is reached, *at which time an internal barrier potential is built up.*

第三种情况：which 在从句中做介词宾语。

Example 8：在分析各类型的电子线路时，首先介绍采用固态器件的那些电路，然后简要地介绍一下采用电子管的电路。

译文：In the analysis of the various types of electronic circuits those using solid-state devices are presented first, *after which a shorter explanation of circuits using electron tubes is presented.*

Example 9：我们使分子比与原子比相等，由此直接得到了最简分子式。

译文：We equate the mole ratio to the atom ratio, *from which the simplest formula follows directly.*

练习：

(1) 制成这台机器的物质是铁。
(2) 存在许多自由电子的物质称为导体。
(3) 我们必须增加可用表的电流范围。
(4) 磁铁能在使其运动的近距离导线中产生电流。
(5) 铁是我们最熟悉的金属之一。
(6) 必定存在一个最佳负载电阻，此时 R 中的功率为最大值。
(7) 要做的第一件事是测出该电阻两端的电压。
(8) 任何热的东西均辐射热量。
(9) 火箭的运行并不取决于空气的存在，这一点在 1916 年就得到了证实。
(10) 大多数绝缘体的电阻率随温度的上升而下降，所以绝缘了的导体必须保持处于低温的状态。

Unit 7　Translation of Adverbial Clauses
（状语从句的翻译）

在汉译英中，有时为了使译文句子结构简明、紧凑，状语可直接翻译成名词性短语，主要有以下几种情况：

1. 时间状语

Example：<u>在过去的五年里</u>，围棋软件虽然进步很大，却从没有重大突破。

译文：<u>The past five years</u> have yielded incremental improvements in Go Programs but no breakthroughs.

2. 目的状语

Example：<u>要让计算机模仿人那样思考</u>牵涉人工智能的核心技术。

译文：<u>The challenge of programming a computer to mimic the thinking process of the human brain</u> goes to the core of artificial intelligence.

3. 条件状语

Example 1：<u>如果研制出更轻的材料</u>，就能生产具有更大"能量密度"的电池。

译文：<u>Research into lighter materials</u> could produce batteries with greater "energy densities".

Example 2：<u>只要把这个设计稍作修改</u>，就可以大大提高该武器的杀伤力。

译文：<u>Slight modification of the design</u> enable the lethality of the weapon to be greatly enhanced.

4. 原因状语

Example：<u>由于改进了组织架构</u>，如20世纪90年代发展的全球定位系统，炮兵在通信方面的弱点达到最小化。

译文：<u>Improved organization structures</u>, such as the development of global positioning systems in the 1990s, have minimized artillery vulnerabilities in terms of communications.

5. 方式状语

Example 1：<u>通过监测降雨动向和其他降水量</u>，科学家就能了解又一大气谜团——风。

译文：<u>Detecting movement of rain and other precipitation</u> tell scientists about another piece of the atmosphere puzzle—wind.

Example 2：<u>由于持久（巡逻）能力、传感能力、数据处理和通信能力得到增强</u>，无人机在防御、情报和民用领域的应用大大得到扩展。

译文：<u>Improved capabilities in persistence (loitering), sensors, data processing and communications</u> have broadened the use of drones in both defense, intelligence and civil roles.

此外，将汉语中的状语直接翻译成英语状语从句时一般只要注意正确选择从属连词。而由于英文的同义连词较多，因此使用时需要注意同义连词的适用性差异，下面列出了几组常见的易混淆使用的同义连词。

PART III　Techniques of EST C-E Translation（科技英语汉译英翻译技巧）

1. 表示原因的连词

常见的有下面几个，其表示原因的语气由强至弱的排列如下：

1) because（= in that）：它表示的原因构成了句子的最主要部分，because 引导的从句一般放在主句后面，需要强调也可放在主句之前。汉语可说"因为……所以……"，但英语的 because 是不能与 so 相呼应的。

Example 1：旋转的物体具有动能，因为其所有的组成微粒都处于运动状态。

译文：A rotating body processes kinetic energy *because its constituent particles are in motion.*

2) since（= in that）：它表示的原因已为人们所知，或不如句子的其余部分重要；since 所强调的是主句的内容，与 as 的不同之处在于它表示稍加分析之后而推断出来的原因。从句一般放在主句前（科技文中也常放在主句后）。

Example 2：既然 k 和 m 均为常数，所以 k/m 比值是恒定的。

译文：*Since k and m are both constants*, the ratio k/m is constant.

Example 3：显然不可能存在电流流动，因为就根本没有电路。

译文：Obviously, no current can flow *since there is no circuit.*

3) as（= now that）：其用法与 since 类同，但是 as 引导从句时语气比 because 弱，as 从句常出现在主句之前，表示比较明显的、已为大家所知的原因。

Example 4：由于空气具有重量，所以它对处于其中的任何物体都要施加一个力。

译文：*As air has weight*, it exerts force on any object immersed in it.

Example 5：由于我们已讨论了联立方程组的图解含义及画线的方法，我们现在能求出线性方程组的图解了。

译文：*Now that we have discussed the meaning of a graphical solution of a system of simultaneous equations and the method of plotting a line*, we are in a position to find graphical solutions of systems of linear equations.

4) for：由它引出的句子使人觉得所述的理由只是一种补充说明而已，它是一个等立连词，引出的句子为并列句而不是状语从句，它只能放在前一句的后面，偶尔也有单独成为一句的。

Example 6：在前面几章我们并没有使用三角函数、反三角函数或对数函数，因为其各自的导数均为一种特殊的形式。

译文：In previous chapters we did not use the trigonometric, inverse trigonometric, exponential, or logarithmic functions, *for the derivative of each of these is a special form.*

2. 表示"当……时候"的连词

1) when：它引出的从句表示的动作往往为一点，而这时主句的动作也多表示为一点。

Example 1：当火箭从地球表面发射时，其发动机的推力必须超过其重量以使它升离地面。

译文：*When a rocket is launched from the earth's surface*, the thrust of its engines must exceed its weight for it to rise from the ground.

2) while：它引出的从句往往采用进行时态，从句中的动作一般持续一段时间，而主句的动作则多表示一点或一种状态。

Example 2：物体转动时能够固定在某一地点。

译文：It is possible for a body to remain in one place *while it is rotating*.

Example 3：当晶体管截止时，交流负载线是水平的。

译文：*While the transistor is off*, the AC load line is horizontal.

3) as：它引导的从句中的动作一般与主句的动作均表示持续一段时间，它们同时发生。

Example 4：当无线电波沿地球表面传播时，要损失掉一部分能量。

译文：*As radio waves travel along the surface of the earth*, part of its energy will be lost.

3. 表示"虽然"的连词

在科技文中常用的有以下几个：

1) although：它一般用于正式的场合，同时可用于各种文体。

Example 1：前面所讲的内容为这个定理提供了基础，虽然不能当作一种证明。

译文：The foregoing provides a basis for this theorem, *although it cannot be considered as a roof*.

2) though：它一般用于非正式的口语或书面语中。该从句可采用特殊词序（在科技文中一般是做表语的形容词放在 though 之前）。

Example 2：该定律虽然重要，但在实践中很少使用。

译文：*Important though this law is*, it is seldom used in practice.

3) as：用于正式的文体中，从句一定要采用特殊词序（在科技文中主要是做表语的形容词放在 as 之前）。

Example 3：电子虽小，但它们在形成电流方面起了重要作用。

译文：*Small as electrons are*, they play an important role in the formation of electric current.

4) while：它引导的从句侧重于对比；当主句和从句的句型相同时，一般把 while 译成"而"字（若这时 while 从句放在主句前的话，则把"而"字译在主句之前）。

Example 4：这时动能趋于无穷大，而势能则趋于最小值。

译文：At this time the kinetic energy approaches infinity, *while the potential energy approaches the minimum*.

Example 5：输入 A 变成了低电位而输入 B 则保持高电位。

译文：Input A goes low *while input B remains high*.

练习：

(1) 计算机产业在过去 10 年里取得了巨大的进展。
(2) 由于进给范围很广，本机床可适于许多种工件的加工。
(3) 汽缸内蒸汽膨胀时，部分蒸汽冷凝。
(4) 因为物体的质量不变，所以力增加的结果就会是速度增加。
(5) 激光可用作外科手术器械，因为光束可立刻切开细胞组织。
(6) 太阳和砂都是物体，因为它们都是由一定数量的物质组成的。
(7) 由于铝具有很高的导热性和导电性，所以它在工业上获得了广泛的应用。
(8) 只要在一个铜导体的两端加上零点几伏的电压就会产生一个相当大的电流。
(9) 请调低收音机，有人还在办公室工作。
(10) 还是有效果产生，即使不总是容易被发现。

Unit 8   Omission in Sentence Translation
（句子翻译中的省略）

在汉译英过程中，原文中有些词语从英语表达的习惯、句子结构和修辞特点来看，往往都是多余的，可以或必须略去。运用省略译法一般可遵循这样的原则，即省略的词语必须是：在译文中看来是可有可无或是多余的；或其意思已经包含在上下文里；或其含义在译文中是不言而喻的。汉英翻译时，省略法运用的场合除了动词和名词的省略，还有以下几种情况：

1. 事理逻辑省略

科技文章中有些阐明事理、探讨问题和描述过程的词句和段落，其中有些词语在汉语表达中必须有，才符合事理逻辑，才顺理成章，但在英语中往往也由于采用不同内涵的词汇或不同的表达方式，而不再需要这些话。

Example 1：我们不能完全排除其他行星上有生命<u>存在</u>的可能性。

译文：We cannot definitely rule out the possibility of life on other planets.

Example 2：利用可视电话，不仅可以听到通话人的声音，而且还能看见其<u>人</u>。

译文：With a videophone you not only hear the person you are talking to, but also see him.

Example 3：空气、食物、水和热量是一切生物赖以<u>生存</u>的四个条件。

译文：Air, food, water and heat are four requirements of all living things.

另外，在操作规程里、使用说明书等一类资料中，按照汉语的事理逻辑，经常在"检查"等术语后用"是否""有无""到底是多少"等语气词构成选择语气。其英译却常常是省略这些字样，而以 inspect、check、examine、signal 等动词译作谓语构成肯定语气的句子。

Example 4：从包装箱中取出仪器时，请仔细检查<u>有无</u>损伤之处。

译文：Inspect the instrument carefully for damage when removing it from the protective constrainer.

Example 5：在使用测压器之前，应检查其功能<u>是否</u>正常。

译文：Voltage testers should be checked for correct functioning before use.

2. 语言逻辑省略

由于汉英两种语言中，使用连接词语的习惯不同，所以汉语中有些句子的各个成分之间、有些句子与句子之间，需要采用某些词语才能将层次、条理和关系表达清楚，但在译成英语时，若对等译出反而不符合英语的规范。这时就需要适当省略一些词语，这就是语言逻辑省略。例如，汉语偏正复句里的关联词可以成对出现，而英语主从复合句里的从属连词却只用一个，因此汉译英时要略去一个。此外，最常见的就是省略连接词语。按照汉语的语言逻辑，句子内部的某些成分之间，或在句与句之间，习惯用一些连接词语，如"就""又""然而""然后"等，才能把层次和条理表达清楚，使文字衔接紧密，语义贯通。而译成英语时，这些关联词往往要省去不译才符合行文的习惯。

Example 1：锰是一种灰白色的、<u>又</u>硬<u>又</u>脆的金属。

译文：Manganese is a hard, brittle, grey-white metal.

Example 2：理论固然重要，实践尤其重要。

译文：Theory is something but practice is everything.

3. 修辞省略

由于汉语和英语两种语言各自的修辞方式有很大的区别，表达方式也各有千秋，因此，翻译时就得顺着原文的意思，灵活、恰当地略去原文中的一些文字。省略解释性和概括性词语就是所谓修辞省略的两种情形。

（1）省略解释性词语

Example 1：这就是汽轮机的工作原理。

译文：This is the principle of the steam turbine.

Example 2：反应时间和发射速度可以互相补偿，至少在一定程度上是如此。

译文：Reaction time and rate of fire compensate for each other, at least to a certain extent.

（2）省略概括性词语

Example 1：声音的频率、波长和速度三者是密切相关的。

译文：The frequency, wave length, and speed of sound are closely related.

Example 2：真空管的五大主要功能是：整流、放大、震荡、调制和检波。

译文：The principle functions that may be performed by vacuum tubes are rectification, amplification, oscillation, modulation and detection.

4. 省略重复成分

汉语中为了讲究句子的平衡、气势或韵调，常常使用排比、对仗和重复等手段。因此汉语句子中，一些词或短语重复使用的现象和结构类似、含义相同的几个词组连用的现象相当普遍。而英语不然。科技英语文章为使语言简洁、表达有力，避免重复，或因习惯或语法结构上的要求，却常常省略句子中的一个或几个成分。这种省略有些已成为英语语言运用中的一种习惯。因此，汉译英时，原文中含义重复的词语往往只译出其中的一个，其他的则略去不译。这种情形在平行结构和偏正结构中都有。

（1）并列句中重复成分的省略

汉语中，在有连词或没有连词的并列句中，第二个分句里常会重复出现与第一分句相同的成分。而在英语中，这种重复成分大多是全部或部分地被省略掉的。因此，翻译这类句子时，需要把这些重复成分全部或部分地略去。

Example 1：质子带阳电，电子带阴电，而中子既不带阳电，也不带阴电。

译文：A proton has a positive charge and an electron a negative charge, but a neutron has neither.

Example 2：增大时的加速度为正，速度减小时加速度为负。

译文：The acceleration is positive if the speed is increased, negative if the speed is decreased.

Example 3：10 的对数等于 1，100 的对数等于 2，0.001 的对数等于 -3。

译文：The logarithm of 10 is 1; of 100, 2; of 0.001, -3.

（2）偏正复句中重复成分的省略

偏正复句由正句和偏句两部分组成。正句与偏句之间的关系是不平等的，有主有次，有正有偏。正句承担了复句的基本意思，是基本的、是主要的；偏句修饰或限制主句，是辅助的、次要的。偏正复句中的省略现象比较复杂，有时难以判断所重复的成分，也难以机械地把重复成分略去。因此，对这类句子的情况应认真分析。

Example 1：远距离送电需要高压，而安全用电则需要低压。

译文：High voltage is necessary for long transmission line while low voltage for safe use.

Example 2：当加工工件时，工件转动，但刀具不动，刀具只是在工件转动时进刀。

译文：When a workpiece is being machined, the workpiece turns, but cutting tools doesn't, it only feeds as the workpiece is being turned.

在比较状语从句中省略重复的词语更是常有的现象。

Example 3：从水中分离氢要比从酸中分离氢困难。

译文：It is difficult to free hydrogen from water than from acids.

Example 4：若是判断一个物体是否会在水中浮起，就得知道该物体的密度比水大还是比水小。要是比水大，该物体下沉，要是比水小，就会浮起。

译文：When trying to decide whether an object will float in water, you need to know whether its density is greater or less than that of water. If it is greater, the object will sink. If less, the object will float.

(3) 含义相同重复词语翻译成共有的句子成分

汉语中为了把内容的搭配或修饰关系表达清楚，常常重复使用相同的成分。例如，汉语习惯上说"分析问题、解决问题"（即重复说出相同的宾语"问题"），而不把宾语"问题"当作动词"分析"和"解决"的共有成分，说成"分析、解决问题"，但在英语中却不是这样的。英译是把 problems 译作 to analyze and solve 的共有宾语。说成 to analyze and solve problems，而不是重复 problems 说成 to analyze problems and solve problems。由此可以看出，汉译英时，对于重复使用的相同成分，只需译出其中的一个，其他的则一概予以省略。

1) 将重复的宾语译成共有宾语。

Example 1：这种 X 射线照片对诊断多种疾病，或者说找出多种疾病的病因很有用。

译文：These X-pictures are useful in diagnosing, or finding the cause of many illnesses.

Example 2：光线射入水中，或从水中射出时，其方向略有变化。

译文：When light rays pass into or out of water, they change their direction slightly.

2) 将重复的动词译成共有动词谓语

Example 3：弗莱明证实，青霉素不仅能杀灭葡萄球菌，而且也能杀灭其他种类的细菌。

译文：Fleming proved that penicillin could kill not only staphylococcus bacteria but other kinds also.

3) 将重复的定语或中心语译成共有定语或共有被修饰语

Example 4：如我们会看到这种结构的标题：

压力传感器和压力扫描阀 Pressure Transducers and Scanner Valves

半导体存储器和渡线存储器 Semiconductor and Plated-Wire Memories

Example 5：金属热处理最常用的是煤气灶、油炉和电炉。

译文：Coal, gas, and electric furnaces are most commonly used for heat treating metal.

Example 6：在运输领域，电气工程师当前正从事于电动汽车、电动火车、电动公共汽车和电动轮船的研制工作。

译文：In the field of transportation, electrical engineers are currently engaged in developing the electrical automobile, train, bus, and ship.

4）重复的中心语译成 of 短语后置定语的共有被修饰语。

Example 7：原子核的电荷和电子的电荷相等，所以原子不带电。

译文：The charges of nucleus and electrons are equal so that the atom is electrically neutral.

Example 8：用来加工塑料制品的机器同那些用来加工木制品、金属制品与其他天然材料制品的机器，是十分不同的。

译文：Machines for making plastic objects are very different from used for manufacturing articles of wood or metal or other nature materials.

5）将重复的形容词或介词短语定语译作共有定语

Example 9：李教授写了几篇论文还有几本书，都是关于理论物理的。

译文：Professor Li has written several articles and books on high energy physics.

Example 10：化工厂是生产化工产品即人造液体、人造气体和人造固体材料的工厂。

译文：Chemical works are factories which produces chemicals—man-made liquids, gases, and solid materials.

练习：

(1) 他对所发现的不同的动植物数量之多感到惊讶。

(2) 仪表面板上的指示灯可向操作者显示出控制器的工作是否正常。

(3) 安培、欧姆、伏特这三个单位分别是以三位科学家的姓氏命名的。

(4) 在匀速运动方程式中，表示距离的符号是 d，表示速度的是 v，表示时间的是 t。

(5) 也已证实，这种塑料比任何一种合金都要更适合于这类用途，而且价格也比任何一种合金都低得多。

(6) 在鉴定试验期间和试验结束，又对原设计做了进一步的改进。

(7) 目前使用的大多数连续自动工作机床采用电-气动技术或电-液压技术。

(8) 不要在过冷、过热、灰尘过重或湿度过大的环境中使用该微型计算机。

(9) 条条规则都有例外。

(10) 物质受热膨胀的原因人人皆知。

附 录

附录 I Glossary for Electronic Engineering
（电子工程专业技术词汇）

absorption law 吸收律（布尔代数）
acceptor impurity 受体杂质
AC coupling 交流耦合
acoustic noise 声音噪声
actuating force 驱动力
admittance 导纳
adjustment pot 调整电位器
ammeter 安培计
ampacity 载流量，安培容量
amperite（ballast）tube 镇流管
amplitude 幅度
am receiver 调幅接收机
and gate 与门
anisotropic 各向异性的，非均质的
anode 阳极
astable multivibrator 无稳态多谐振荡器
asynchronous logic 异步逻辑
attenuator pad 衰减器
attraction 吸引
audio amplifier 音频放大器
avalanche effect 雪崩效应
axial lead 轴向引线
azimuth 方位角
background noise 背景噪声
back resistance 反向电阻
balanced phase detector 平衡鉴相器
bandwidth 带宽
barretter 镇流电阻器
baud 波特（传码率的单位）
beam 光束
beat-frequency oscillator 差频振荡器

bias current 偏置电流
bistable multivibrator 双稳态多谐振荡器
bleeder current 漏电流
blocking oscillator 阻塞振荡器
boolean algebra 布尔代数
breakdown voltage 击穿电压
bridge circuit 电桥电路
brightness control 亮度控制
buffered delay line 缓冲延迟线
built-in test equipment（BITE） 内置测试装置
bypass capacitor 旁路电容
capacitive coupling 电容耦合
cathode bias 阴极偏置
cell 单元，电池
characteristic impedance 特征阻抗
charge cycle 充电周期
circuit breaker 断路器，断路开关
clamper 嵌位器
clipper 嵌位器
coaxial line 同轴线
coefficient of coupling 耦合系数
coherent oscillator 相干振荡器
coil （轴）线
combinational logic 组合逻辑
common-mode interference 共模干扰
commutative law 交换律
commutator 换向器
compander 压扩器
comparator 比较器
compensating windings 补偿线圈

complex wave 复合波
compressor 压缩器
conductance 电导
conductivity 导电性
connector adaptor 连接适配器
constant current source 恒流源
converter 转换器
corona 电晕放电
critical frequency 临界频率，截止频率
cathode ray tube 阴极射线管
crystal oscillator 晶体振荡器
current standing-wave ratio（CSWR） 电流驻波比（即最大和最小电流之比）
cutoff frequency 截止频率
decoupling capacitor 去耦合电容器
deflection coils 偏转线圈
delayed sweep 延时扫描
DeMorgan's theorem 德-摩根定理
demultiplexing 多路分解
depletion region 耗散区
detection 检测
dielectric 绝缘体，电介质
differential amplifier 差分放大器
diffraction 衍射
dipole 偶极
directional coupler 定向耦合器
discrete component 分离元件
discriminator 鉴相器
distilled water 蒸馏水
distortion 失真
distributed constant 分布常数
distributive law 分配律
domain theory 磁畴理论
doping 掺杂
dropout 信号丢失
drive-by-wire 线控
dropout voltage 输入输出电压差
dual-gate mosfet 双栅场效应晶体管
dual in-line package（DIP） 双列直插式组件
dummy load 假负载，虚负载
duplex circuit 双向（工）电路
duty cycle 占空比
dynamic range 动态范围
emitter coupled logic 发射极耦合逻辑
eddy current 涡流
edge triggered flip flop 边缘触发触发器
electrical charge 电荷
electrical zero 零电位
electrolyte 电解质
electromagnetism 电磁学
electromotive force（EMF） 电动势
electronic frequency counter 电子频率计数器
electronics 电子学
electrostatic 静电
emergency power 应急电源
equivalent resistance 等效电阻
excitation voltage 激励电压
exclusive-or（XOR） 异或
exclusive-nor（XNOR） 同或
fast-time-constant circuit 短时定数回路
fault indicator 故障指示器
ferromagnetic material 铁磁材料
fidelity 保真度
field-effect transistor（FET） 场效应晶体管
field-programmable logic array（FPLA） 现场可编程逻辑阵列
field strength 电场强度
final power amplifier（FPA） 末级功率放大器
fixed resistor 固定电阻器
floating input 悬浮输入端
flywheel effect 飞轮效应（即振荡回路能够利用存储的能量工作的现象）
forbidden band 禁带
forward resistance 前向电阻
free-space loss 空间损耗
frequency synthesiser 频率合成器

full-duplex circuit　全双工电路
full-wave rectifier　全波整流器
galvanometer　电流计
ganged tuning　联动调谐
gating　选通
grid current　栅电流
grid-leak bias　栅漏偏置
ground waves　地波
half-power point　半功率点
harmonic distortion　谐波失真
hertz（Hz）　赫兹（频率单位）
high frequency compensation　高频补偿
hold time　维持时间
hole flow　空穴流
horizontally polarized　水平极化的
horseshoe magnet　U形磁体
hot carrier　热载流子
hot-plug slot　热插槽
hot swap　热插拔
hybrid circuit　混合电路
hysteresis　磁滞
idle-channel noise　闲置通道噪声
idempotent law　幂等定律
impedance matching　阻抗匹配
in-circuit meter　在线仪表
indirectly heated cathode　旁热式阴极，间接加热阴极
induced charge　感应电荷
induced electromotive force　感应电动势
input impedance　输入电阻
input/output　输入/输出
insertion loss　插入损耗
instantaneous amplitude　瞬时振幅
instrumentation amplifier　测量放大器
insulated gate field-effect transistor（IGFET）　绝缘门场效应晶体管
insulation　绝缘
integrated circuit（IC）　集成电路
integrator　积分器

interelectrode capacitance　极间电容
intermediate power amplifier　中间功率放大器
intrinsic failure rate　固有故障率
inverter　反相器
inverting amplifier　反相放大器
ionize　离子化
ionosphere　电离层
isochronous distortion　同步失真
isolator　隔离器，绝缘体
jack　插座
joint test action group　调试接口
johnson noise　热噪声
joule　焦耳，能量的单位
joystick　操纵杆
junction　结点
junction transistor　结型晶体管
karnaugh map　卡诺图
keep-alive voltage　激励电压
kinetic energy　动能
klystron power amplifier　调速管功率放大器
latch　锁存器
LC capacitor-input filter　LC电容输入滤波器
lead inductance　导线电感
lead sheath　铅包护皮，铅护套
leakage current　漏电流
light-emitting diode（LED）　发光二极管
left-hand rule　左手定则
linear impedance　线性阻抗
line of force　（电或者磁）力线
lin-log amplifier　线性对数放大器
liquid crystal display　液晶显示器
lithium-ion battery　锂电池
loading coil　负载线圈
loading effect　负载效应
local action　自放电
local loop　本地回路
logarithmic receiver　对数接收机
logical block addressing　逻辑块寻址
logic circuit　逻辑电路

loose coupling　疏耦合
magnetic amplifier　磁放大器
magnitude comparator　幅度比较器
master oscillator　主振荡器
maximum power dissipation　最大功率损耗
metal-oxide semiconductor field-effect transistor（MOSFET）　金属氧化物半导体场效应晶体管
metal oxide varistor　压敏电阻
meter shunt　仪表分流
mho　姆欧（姆电导单位，欧姆的倒数）
microcircuit　微型电路，超小型集成电路
microelectronics　微电子学
millimeter wave　毫米波［波长从0.1mm到1mm（频率从300GHz到3000GHz）］
minimum discernible signal（MDS）　最小可识别信号
minority carrier　少数载流子
minority carrier current　少数载流电流
minor lobe　副瓣，旁瓣
mixer　混频器，混合器
modular circuitry　模块化电路
modulated oscillator　可调振荡器
monostable multivibrator　单稳态多谐振荡器
moving coil speaker　动圈式扬声器，移动线圈扬声器
multiconductor　多触点，多导体
multicoupler　多路耦合器
multielectrode tube　多（电）极（电子）管
multiphase　多相
multiplication factor　倍增因数
multispeed synchro system　多速率同步随动系统
multi-turn potentiometer　多圈电位器
multi-unit tube　多组管
multivibrator modulator　多谐振荡器式调制器
nand　与非门
N-channel fet　N型场效应晶体管

near synchronous orbit　近同步轨道
negative electrode　负极
negative-resistance element　负电阻部件
nesting　嵌套
neutralization　中和，中和作用，中立状态
node　节点，叉点
noise margin　噪声容限，杂讯容限，噪声安全系数
nondegenerative parametric amplifier　非退化参量放大器
non-inverting amplifier　非反相放大器，同相放大器
nonlinear impedance　非线性阻抗
nonresonant line　非共振线
nor gate　或非门
off-line test equipment　离线测试设备
off set　偏移
ohmmeter　欧姆计，电阻表
Ohm's law　欧姆定律
open collector　（OC）集电极开路，集电极开路输出门
operational amplifier　运算放大器
optimum working frequency　最佳工作频率
order-wire circuit　命令线电路
or gate　或门
oscillator　振荡器
output rating　输出率，额定输出功率
overcurrent　过载电流，过量电流
parallel circuit　并联电路
passive filter　无源滤波器
path loss　路径损耗
phase-locked loop（PLL）　锁相环
phase splitter　分相器
phototransistor　光电晶体管
picopref　兆分之一（相当于10^{-12}）
piezoelectric effect　压电效应
pi filter　π型滤波器
pigtail　抽头
pin diode　点触型二极管

plate dissipation 板极损耗
plenum cable 夹层电缆
point-of-load (POL) 荷载点
pole piece 极片，磁极片
positive logic 正逻辑
potentiometer 电位计，分压计
preemphasis 预矫，预修正，预加重
primary cell 原电池
programmable logic array (PLA) 可编程序逻辑阵列
programmable logic device (PLD) 可编程逻辑电路
propagation time delay 传播时间延迟
prototype 原型，标准，模范
pseudorandom number generator 伪随机数发生器
pulse-response 脉冲响应
push-pull amplifier 推挽放大器
quality of sound 音质
quartz oscillator 石英振荡器，水晶振子
quiescent current 无信号电流，静态电流
quiescent state 静态，静止状态
R2RladderR/2R 阶梯网络
race condition 竞争条件，竞态条件
raceway 电缆管道
radiation resistance 辐射抗性，耐辐射性，辐射电阻
radix point 小数点
rated output power 功率，输出功率，额定功率
ratio 比率，比例
reactance 电抗，感应抵抗
recovery time 恢复时间
rectifier 整流器
reference line 参考线，基准线
reflector 反射物，反射镜，反射器
refractive index 折射率，折光率
regenerative feedback 再生反馈，正反馈
regulator 调整者，校准器

relative bearing 相对方位
reliability 可靠性
reluctance 磁阻
repeater 中继器，转发器
residual magnetism 残余磁性
resolution 分辨率，清晰度
resultant magnetic field 合成磁场，复合磁场
return loss 回波损耗
reverse bias 反向偏压
RF amplifier 射频放大器
rheostat 变阻器
ribbon cable 带状电缆
right angle connector 直角连接器
rigid coaxial line 硬同轴线
ripple factor 脉动系数，纹波因数，纹波系数
RLC circuit RLC 电路，阻容网络
root-raised-cosine filter 根余弦滤波器
rotary switch 旋转开关
sample and hold 采样和保持，采样及保持电路
saw tooth 锯齿，锯齿形
Schmitt trigger 施密特触发器
Schottky diode 肖特基二极管
silicon controlled rectifier 可控硅整流器
secondary emission 次级发射，二次发射
selectivity 选择性
self-bias 自给偏压
sense line 感应线
sensor 传感器
series-fed oscillator 串联供电振荡器
shaping circuit 整形电路
sharp-cutoff tube 锐截止管
set-reset flip flop 置位复位触发器
shielded cable 屏蔽电缆，屏蔽线
shift register 移位寄存器
short circuit 短路
shot noise 散粒噪声，散射噪声

shunt-diode detector 旁路二极管检测器
shunt-fed oscillator 分路馈电式振荡器
single inline package 单列直插封装，单列直插式组件
single line diagram 单线图
sink current 反向电流
skin effect 集肤效应
skip distance 跳跃距离
slew rate 转换速率，转换速度
specific resistance 电阻率，电阻系数
spectrum analysis 频谱分析
speed-up capacitor 加速电容，加速电容器
square wave 方波
squirrel-cage windings 鼠笼式绕组
stability 稳定性
stagger tuning 串联调谐，交错调谐
standing wave 驻波
step-up transformer 升压变压器，升压器
stranded conductor 绞合导体，绞合导线
summing circuit 加法电路
suppression 抑制
swamping resistor 扩量程电阻器
synthesizer 合成器
tank circuit 储能电路，振荡回路
tapped resistor 抽头式电阻器
temperature coefficient 温度系数
thermistor 热敏电阻，电热调节器
thermostatic switch 恒温开关
thyristor 半导体闸流管
time base 时基，扫描基线，扫描发生器
time delay relay 延时继电器
time-domain reflectometer 时域反射计
timing skew 时序偏差
transconductance 跨导
transistor 晶体管
trigger generator 触发脉冲发生器
true power 有效功率
truth table 真值表
tuned circuit 调谐电路

tunnel diode 隧道二极管
turn 匝，转，转动
tweeter 高音用扩音器，高频扬声器
twinax 屏蔽双导线馈电线
ultra extended graphics array (UXGA) 极速扩展图形阵列
ultrasonic 超声波，超音速的
unbalanced crystal mixer 非平衡晶体管混频器
uniform transmission line 均匀传输线
unijunction transistor (UJT) 单结晶体管
unit impulse 单位脉冲
untuned line 未调谐线路
vacuum tube 真空管
varactor 变抗器，变容二极管
variable capacitance diode 变容二极管
varistor 压敏电阻，变阻器
Venn diagram 维恩图
vertically polarized 垂直极化
virtual ground 虚地，虚拟接地
visible spectrum 可见光谱
voice coil 音圈
voltage controlled oscillator 压控振荡器
voltage spike 突增的瞬间电压
voltage standing wave ratio 电压驻波比
voltage transient 电压瞬变
voltaic cell 伏打电池，原电池
voltmeter 伏特计，电压表
volt-ohm meter (VOM) 伏-欧表
wafer switch 晶片开关
watchdog timer 看门狗定时器
watt 瓦特（功率单位）
wattmeter 瓦特计，功率计
wavelength 波长
Weber's theory 韦伯理论
wheatstone bridge 单臂电桥，惠斯通电桥
white noise 白噪声
wire bundle 导线束交通，导线束，钢丝束
wiring diagram 接线图，布线图

working voltage 工作电压
X-ray emission X 射线发射
xy-cut crystal xy 截面晶体
y-amplifier 示波器的垂直放大器
y-tilt y 轴倾角
zener diode 齐纳二极管

zero adjustment 零位调整，零点调整
zero insertion force socket 无插拔力插座，零插入阻力插座
zig zag package 锯齿形封装插座
zoom lens 变焦镜头

附录 II　Glossary for Communication Engineering
（通信工程专业技术词汇）

absent service 缺席用户服务
access barred signal 接入拒绝信号
access mask 访问掩码
accessory kit 附件箱
accounting module 计费模块
acknowledge fault alarm 确认故障告警
adaptive antenna array 自适应天线阵列
addressing 寻址
adjacency 邻接
adjustable range 可调范围
admitted deviations 可接受的偏差
advertise 发布，传送
aging element parameter 老化因子参数
alarm buzzer 报警蜂鸣器
alignment error rate monitoring 定位差错率监视
alternate mark inversion encoding AMI 编码
analog front end 模拟前端
answer file 应答文件
anti-ghost 消重影，叠影消除
assembly drawing 装配图
associated control channel 随路控制信道
asymmetric key algorithm 非对称密钥算法
asynchronous transmission 异步传输
at the remote end 在远端
ATM switch ATM 交换机
attachment unit interface 连接单元接口

attenuation 光路衰减
audio coder/decoder 语音编/解码器
audio frequency 音频
authoritative restore 授权恢复
auto correlation 自相关
automatic gain control (AGC) 自动增益控制
autosensingn 自适应，自动侦测
auxiliary port 备份口
back report 后向报告
back-to-back connection 背靠背连接
backbone network 骨干网
balanced audio 平衡音频
bandwidth allocation 带宽分配
base station 基站
Baud rate 波特率
be configured to the full capacity 满配置
bearer channel 承载信道
bi-directional 双向的
binary operator 双目运算符，二进制运算符
binary phase shift keying 二相移相键控
binding key 绑定密钥
bit error rate 误码率
bit timing 位定时
bit-oriented 面向比特的
black-and-white image 黑白图像
blackhole route 黑洞路由
boolean value output sensor 开关量传感器

broadband multi-service access system 宽带多业务接入设备
broadcast 广播
brouter 桥路器
buffer 缓冲区，缓冲器，减震器
built-in camera 内置摄像机
bulk encryption 批量加密
bundling trough 绑线槽
burst transmission 突发传输
busy-hour statistics 忙时统计
buzzer interface 蜂鸣器接口
cable removing 拆线
cable TV 有线电视
call bearer control 呼叫承载控制
capability exchange 能力交换
capacity configuration 容量配置
caption function 字幕功能
capture range 捕获范围
carrier detect 载波检测
cascade 层叠，级联
cell breathing 小区呼吸，手机发话范围
cell delay variation tolerance 信元时延抖动容限
cell error ratio 信元错误率
cellular wireless communication networks 蜂窝无线通信网络
chain 链
changeover 转换，切换
channel capacity 信道容量
channel multiplexing 信道复用
character string 字串
check bit 检验位
chromatic dispersion 色度色散
cipher text 密文，密钥
circuit breaker 断路开关，断路器
circuit group blocking 电路群闭塞（的请求）
circuit path list 电路路径列表
clear-backward signal 反向拆线信号
co-processor module 协处理模块

code division multiple access (CDMA) 码分多址
composite video 复合视频
congestion indication 拥塞指示
connecting terminal 接线端子
connectionless network service 无连接网络业务
continuity check 导通检验
controlled-load service 负载控制服务
convergence 会聚，收敛
coupling 联结，耦合
creepage capacitance 漏电电容
criterion value 标准值
cryptographic key 密钥
cyclic code 循环码
data carrier detect 数据载波检测
data encryption 数据加密
deactivate 去激活
dead time 失效时间
decimal notation 十进制
decipher 解密器
demodulation 解调
destination route 目的地址路由
device topology daemon 设备拓扑后台
device view 设备视图
dial pulse 拨号脉冲
dial-on-demand 按需拨号
dictionary attack 字典或攻击
differential frequency hopping 差分跳频
differential phase 微分相位
digital communication multiplexer 数字通信复用器
digital data service 数字数据业务
digital line signalling 数字线路信令
digital mobile communication 数字移动通信
digital signal processing 数字信号处理
digital signalling data link 数字信令数据链路
digital subscriber line access multiplexer 数字用户线接入复接器

direct sequence spread spectrum signal　直接序列扩频信号
disable voice activation　禁止声控
discrimination group number　甄别组号
dispersion compensation　色散补偿
display adapter　显卡
distributed forwarding　分布转发
do not disturb　请勿打扰
DSL access multiplexer　数字用户线接入复用器
dual audio channel　双声道
dual stage dialing　二次拨号
dummy message　伪消息
dust-proof cushion　防尘垫
dynamic cascading　动态级联
e-business　电子商务
echo cancellation　回声消除
edge detection　边缘检测
efficient bandwidth usage　有效带宽使用
electrically-erasable programmable read-only memory　电可擦编程只读存储器
electronic information exchange system　电子信息互换系统
element　码元
emergency communication　应急通信
emulation terminal　仿真终端
encoding and decoding speech　语音编解码
encrypted protection　加密保护
end node　末端节点
endpoint　节点
enhanced color　增强色
enumerated value　枚举值
environment parameter　环境量
equivalent input noise　等效输入噪声
erasable programmable logic device　可擦除可编程逻辑器件
error control　差错控制
error correcting code　纠错码
error rate　差错率

Ethernet　以太网
even charging voltage　均充电压
even parity　偶校验
expansion variable　扩展变量
exponentially　按指数的，按幂值的
extension name　扩展名
external clock reference　外部时钟源
external route　外部路由
fading channel　衰落信道
far end alarm　远端告警
fast convergence　快速收敛
fault tolerance　容错性
fiber access system　光纤接入系统
file transfer　文件传输，文件传送
first harmonic wave　基波，一次谐波
fixed-line phone　固定电话
flash memory　闪速存储器
flow control message　流量控制消息
format conversion　格式转换
forward error correction　前向纠错
forward explicit congestion notification　前向显示拥塞通知
frame rate　帧率
frame relay unit　帧中继单元
frame synchronization control　帧同步控制
frequency-division multiplexing　频分多路复用，频分复用
full duplex digital audio　全双工数字音频
fundamental wave　基波
fuzzy neuron network　模糊神经网络
gateway　网关，关口
Gaussian random noise　高斯随机噪声
general apparatus　通用仪器
generate report　生成报表
gigabyte　千兆字节，吉字节
global call ID　全局呼叫标志号
graphics resolution　图形分辨率
Greenwich Meantime　格林威治标准时间
grounding leading　接地引线

Gaussian minimum shift keying 高斯最小移频键控
hardware flow control 硬件流控
harmonic wave 谐波
header error control 信头差错控制
heat dissipation 散热
hierarchical coding 分级编码
high definition TV 高清晰度电视
high frequency impedance 高频阻抗
high speed signal interface 高速信号接口
high-speed digital link channel 高速数字链路通路
horizontal resolution 水平解像度，水平清晰度，水平分辨率
hybrid digital analog network 数模混合网
hybrid fiber-coaxial 混杂同轴光纤，光纤同轴混合网
hybrid network 混合网络
hyper text markup language 超文本标记语言
hyperlink 超链接
identifier 标志符
illegal intrusion 非法侵入
image compression 图像压缩
image denoising 图像去噪
incoming call barring 呼入禁止，（群内）入呼叫限制
incoming call prohibited 禁止受话
increment value 增量值
independent return transmission networking 独立回传组网
indication command 指示，命令
industrial frequency 工频
input and output control module 输入/输出控制模块
instant echo cancellation 快速回声消除
integrated access device 综合接入设备
integrated digital network 综合数字网
intelligent control center 智能控制中心
intelligent service node 智能业务节点

interleaving and de-interleaving 交织和去交织
intermodulation 互调，相互调制
inter-exchange call 局间呼叫
interoperability 互操作性，互通性
interruption control 中断控制
intra-group call 群内呼叫
inverse discrete cosine transform 反向的离散余弦变换
IP packet IP 数据包
jam-to-signal ratio 干扰信号比
jumper ring 跳线圈
junk mail 垃圾邮件
knowledge based system 知识库系统
large-capacity 大容量
laser communication 激光通信
level conversion 电平转换
lexical analysis 语法分析
light intensity 光强度
lightning strike 雷击
limiter 限幅器
live program 实况转播节目
logic channel identifier 逻辑信道标志
loopback 回送，回路
lossless compression 无损压缩
maintenance configuration mode 维护配置模式
make-break test 通断测试
Manchester encoding 曼彻斯特编码
maximum rate 最大速率
maximum tolerance 最大容限
media access 媒体接入
media attachment unit 介质链接单元
megohmmeter 绝缘电阻表
memory address 内存地址
message distribution 消息分配
microphone inputs 麦克风输入
mismatch 不匹配
misoperation 误操作

modular 模块化的
modulator-demodulator 调制解调器
motion picture expert group 活动图像专家组
movie on demand 影视点播
moving video resolution 活动视频分辨率
multichannel controller 多通道控制器
multi-channel technique 多路计划
multi-level coordination 多级协调
multicarrier 多载波
multimedia 多媒体
multiplex 复用
multiplex frame 组帧
multiuser detection 多用户检测
mute suppression 静音抑制
negative lead 负极连接线
networking mode 组网方式
no parity check 无校验
nonstationary signal 非平稳信号
nonuniform sampling 非均匀采样
normal lighting 日常照明
numeric data 数字数据
object 目标，对象
obstruction 障碍，阻碍，障碍物
odd parity check 奇校验
ohmmeter 电阻表
optical communication 光通信
optical coupling isolation 光耦合隔离
originating point 起源点
out of service 业务中断，失效，停止运行
outband channel 带外信道
packet arrival jitter 包到达抖动
parent channel 父通道
parent domain 父域
parity bit 校验位
parity check 奇偶校验
partial differential equation 偏微分方程
partial matching 不完全匹配
partial packet discard 部分包丢弃
partition knowledge table 分区知识表

passive backplane 无源背板
passive optical transmission 无源光传输
password 密码，口令
password call 密码呼叫
path control and signaling extraction module 通道控制与信令提取模块
PCB Layout PCB 布线
peak cell rate 峰值信元速率
performance analysis 性能分析
peripheral interface 外设接口
permanent virtual path 永久虚通路
phase line 相位线，调整线
phase stability 相位稳定性
phase-locked clock 锁相时钟
plug and play 即插即用
point to point 点对点
pointer 指针，指示器
port 端口
precision charging 精确计费
primary rate adaptation 基群速率适配
pseudo-random number generator 伪随机数发生器
public key 公开密钥，公共密钥
pulse code modulation 脉冲编码调制
quadrature amplitude modulation 正交幅度调制，正交调幅
queueing theory 排队论
random sampling 随机抽样
ranging 排列，测距
Rayleigh fading channel 瑞利衰落信道
real time variable bit rate 实时可变位率
redundancy check 冗余校验
Reed Solomon coding 里德-所罗门编码
region of interest（ROI） 感兴趣区域
registration 配准
regulation factor 调整率
remote bridge 远程网桥
resource allocation 资源调配，资源分配
ring topology 环形网拓扑

sample test　抽测
sampling frequency　采样频率
synthetic aperture radar（SAR）　合成孔径雷达
satellite constellation communication network　卫星星座通信网络
scramble　扰码
secure electronic transaction　安全电子事务
segmentation　分段
self-test　自检，自测
serial data　串行数据
service access point　服务访问点，业务接入点
session based　基于会话的
setting　设置
signal noise ratio　信噪比
signaling route set congestion　信令路由组拥塞
signaling time slot　信令时隙
signed integer　有符号整数
simplex　单一的，单工的
single frequency signaling　单频信令
software development kit　软件开发工具包
sorted user　分类用户
source codec　信源编译码器
special purpose　专用
speckle noise　斑点噪声
speech circuit　话音电路
speech memory　语音存储器
speed (quick) dial　快速呼叫，快速拨号
splitter　分离机，分离器
spread spectrum　扩频
static power consumption　静态功耗
step by step system　步进制，步进方式
step length　步长
streaming player　流媒体播放器
sub-rate board　子速率板
subchannel　子通道，分通道
subscriber cable　用户线，用户电缆

superframe　超帧
switching layer　交换层
switching node　交换节点
synthesized voice　合成声音
tag　页签，制表符
tele-education　远程教学
telemedicine　远程医疗
telephone exchange　电话局，电话交换台
telephone jack　电话插孔
test board　测试板
threshold　门限
time division multiplex　时分复用
time domain equalization　时域均衡
topological view　拓扑视图，拓扑图
touch-tone dialing　按键式拨号，按键拨号
traffic control　流量控制
transaction　交易，事务
transceiver　收发器，收发两用机
transmission disturbance　传输干扰
trellis coded modulation（TCM）　网格编码调制
trunk drive board　中继驱动板
unattended maintenance　无人值守维护
unicast　单点传送，单播
uniconductor　单导体
unreachable route　不可达路由
unreasonable message　不合理的消息
unreliable channel　不可靠通道
unsigned integer　无符号整数
unstructured data transfer　非结构化数据传输
user priority　用户优先级
user defined　使用者定义的，用户自定义的
vacant slot　空槽
vacuum forming　真空成形
value enumeration　值枚举
variable bit rate　可变位率，可变传输率
video cassette recorder（VCR）audio interface　录像机音频接口

videophone 可视电话
videotex 信息传视系统，可视图文系统
view management 视图管理
voice activation 语音激活，话音启动，声控
warranty period 保修期
wavelet image de-noising 小波图像去噪
WDM terminal 波分复用终端
web TV 网络电视
weighted round robin 加权轮叫，权重轮循均衡
whiteboard 电子白板
whole-system consumption 整机功耗
windows open system architecture 窗口开放系统架构
wireless communication system 无线通信系统
wiring closet 布线室
x digital subscriber line x 数字用户线
y-axis Y轴，纵坐标轴
Y-branching device Y形分路器
zero check error 零校验错
zoom in 放大显示

附录 III　Glossary for Automation
（自动化专业技术词汇）

accumulated error 累积误差
actuator sensor interface 执行器传感器接口（ASI 总线）
actuator 驱动器，执行机构
adaptation mechanism 自适应机构
admissible error 容许误差
alternating current（AC） 交流电
amplifying element 放大环节
analog-digital conversion 模数转换
analogue-to-digital converter（ADC） 模数转换器
artificial intelligence 人工智能
assembly workshop 装配车间
asymptotic stability 渐进稳定性
auto pilot 自动驾驶仪
automatic alarm 自动报警器
automatic detection system 自动检测系统
automatic navigator 自动导航仪
automatic production line 自动化生产线
automatic tracking system 自动跟踪系统
backlash characteristic 间隙特性
black box testing approach 黑箱测试法
Bode diagram 博德图
boundary value analysis 边界值分析
butterfly valve 蝶阀
calibration instrument 校准仪器
capacitive displacement transducer 电容式位移传感器
cascade compensation 串联补偿
cascade control 串级控制
central cabinet 中央控制台
characteristic locus 特征轨迹
classical control theory 经典控制理论
closed loop gain 闭环增益
closed loop pole 闭环极点
closed loop transfer function 闭环传递函数
coefficient matrix 系数矩阵
common mode disturbance 共模干扰
common mode rejection ratio（CMRR） 共模抑制比
compatibility 相容性，兼容性
compensating network 补偿网络
computer-aided engineering（CAE） 计算机辅助工程

computer-aided manufacturing (CAM) 计算机辅助制造
computer integrated manufacturing system (CIMS) 计算机集成制造系统
computer integrated process system (CIPS) 计算机集成过程系统
conditionally instability 条件不稳定性
configuration 组态
conservative system 守恒系统
constraint condition 约束条件
continuous duty 连续工作制
continuous oscillation 等幅振荡
continuous system 连续系统
control accuracy 控制精度
control action 控制动作
control algorithm 控制算法
control block diagram 控制方框图
control cabinet 控制柜
control panel 控制屏，控制盘
control plant 控制对象，被控对象
control station (CS) 控制站
control system synthesis 控制系统综合
controllability 可控性
controllability index 可控指数
controllable canonical form 可控规范型
controlled variable 被控变量
controller area network 控制器区域网络（CAN 总线）
controlling instrument 控制仪表
coordination strategy 协调策略
coordinator 协调器
corner frequency 转折频率
correcting element 校正元件
correlation coefficient 相关系数
critical damping 临界阻尼
critical stability 临界稳定性
cross-over frequency 穿越频率，交越频率
cut-off frequency 截止频率
cybernetics 控制论

cyclic remote control 循环遥控
damped oscillation 阻尼振荡
damper 阻尼器
damping ratio 阻尼比
data acquisition 数据采集
data preprocessing 数据预处理
data processor 数据处理器
data terminal equipment (DTE) 数据终端设备
dead time plant 大时滞对象
dead zone 死区
decision support system 决策支持系统
decoupling control 解耦控制
decoupling parameter 解耦参数
deductive-inductive hybrid modeling method 演绎与归纳混合建模法
delay control 延迟控制
delay time 滞后时间
describing function 描述函数
desired value 希望值
detectability 可测性
detection system 检测系统
deviation alarm 偏差报警器
diagnostic model 诊断模型
difference equation 差分方程
differential dynamical system 微分动力学系统
differential pressure controller 差压控制器
differential pressure level meter 差压液位计
differential pressure transmitter 差压变送器
digital control system 数字控制系统
digital-to-analogue converter (DAC) 数模转换器
direct current (DC) 直流电
direct digital control system (DDC) 直接数字控制系统
discrete control system 离散控制系统
discoordination 失协调
discretecontrol system 离散控制系统

discrete event　离散事件
discriminant function　判别函数
displacement transducer　位移传感器
displacement vibration amplitude transducer　位移振幅传感器
distributed control system（DCS）　集散控制系统
distributed parameter control system　分布参数控制系统
disturbance compensation　扰动补偿
disturbance rejection　干扰抑制
diversity　多样性
divisibility　可分性
dominant pole　主导极点
dynamic accuracy　动态精度
dynamic characteristic　动态特性
dynamic deviation　动态偏差
dynamic error coefficient　动态误差系数
dynamic system　动态系统
dynamic data exchange（DDE）　动态数据交换
eddy current thickness meter　电涡流厚度计
electric actuator　电动执行机构
electric conductance levelmeter　电导液位计
electric drive control gear　电动传动控制设备
electric pneumatic converter　电-气转换器
electrohydraulic servo vale　电液伺服阀
electromagnetic flow transducer　电磁流量传感器
electronic batching scale　电子配料秤
electronic belt conveyor scale　电子皮带秤
emergency stop　异常停止
electrostatic coupling　静电耦合
engineer station（ES）　工程师站
equilibrium point　平衡点
error correction device　纠错装置
expected characteristic　希望特性
expert system　专家系统
external disturbance　外扰
extreme value　极值
failure diagnosis/fault diagnosis　故障诊断
fault recognition　故障识别
fault-tolerance control　容错控制
feasibility study　可行性研究
feature detection　特征检测
feature extraction　特征提取
feedback compensation　反馈补偿
feedback correction　反馈校正
feedback system　反馈系统
feedforward control　前馈控制
feedforward path　前馈通路
field bus　现场总线
field bus control system　现场总线控制系统
Fieldbus Foundation（FF）　现场总线基金会
final-value theorem　终值定理
fixed set point control　定值控制
flexible manufacturing system（FMS）　柔性制造系统
float levelmeter　浮子液位计
flow controller　流量控制器
flow sensor/transducer　流量传感器
flow transmitter　流量变送器
flowmeter　流量计
forward path　正向通道
frequency converter　变频器
frequency domain analysis　频域分析
frequency regulator　频率调节器
full order observer　全阶观测器
full scale response　满量程响应
full-automatic control　全自动控制
function block　功能块
functional decomposition　功能分解
fuzzy control　模糊控制
fuzzy logic　模糊逻辑
gain margin　增益裕度
generalized least squares estimation　广义最小二乘估计

global asymptotic stability　全局渐进稳定性
global optimum　全局最优
globe valve　球形阀
Hall displacement transducer　霍尔式位移传感器
hardware-in-the-loop simulation　半实物仿真
heuristic inference　启发式推理
hierarchical control　递阶控制
hierarchical planning　递阶规划
highway addressable remote transducer　可寻址远程变送器数据通路（HART总线）
ideal value　理想值
identifiability　可辨识性
image recognition　图像识别
impulse function　冲击函数，脉冲函数
impulse response　脉冲响应
incremental motion control　增量运动控制
inductive force transducer　电感式位移传感器
industrial automation　工业自动化
industrial personal computer（IPC）　工业控制计算机
inference engine　推理机
information acquisition　信息采集
infrared gas analyzer　红外线气体分析仪
inherent nonlinearity　固有非线性
initial deviation　初始偏差
initial theorem　初值定理
input output interface　输入输出接口
instability　不稳定性
integral action time constant　积分作用时间常数
integral action time　积分作用时间
integral of absolute value of error criterion　绝对误差积分准则
integral of squared error criterion　平方误差积分准则
integration instrument　积算仪器
intelligent control　智能控制
intelligent decision support system　智能决策支持系统
intelligent terminal　智能终端
interacted system　互联系统，关联系统
intermittent duty　断续工作制
internal disturbance　内部干扰
inverse Nyquist diagram　逆奈奎斯特图
Kalman-Bucy filer　卡尔曼-布西滤波器
knowledge acquisition　知识获取
knowledge representation　知识表达
lag-lead compensation　滞后超前补偿
Laplace transformation　拉普拉斯变换
large scale system　大系统
least squares criterion　最小二乘准则
level switch　物位开关
levelmeter　物位计
Liapunov's stability criteria　李雅普诺夫稳定判据
limit cycle　极限环
linear motion valve　直行程阀
linear programming　线性规划
linear quadratic regulator problem（LQR）　线性二次调节器问题
linear time-invariant control system　线性定常系统
local asymptotic stability　局部渐近稳定性
local operating network　局部操作网络（LON总线）
local optimum　局部最优
log magnitude-phase diagram　对数幅相图
lumped parameter model　集总参数模型
Lyapunov theorem of asymptotic stability　李雅普诺夫渐近稳定性定理
magnitude margin　幅值裕度
magnitude-frequency characteristic　幅频特性
man machine interface（MMI）　人机接口
manipulation variable　控制量，操作量
manipulator　操纵器，操作者机械手
man-machine coordination　人机协调

manual operation   手动操作
manufactory management computer（MMC）   生产管理计算机
manufacturing automation   制造自动化
master station   主站
matching criterion   匹配准则
maximum likelihood estimation   最大似然估计
maximum overshoot   最大超调量
maximum principle   极大值原理
mean time between failures（MTBF）   平均故障间隔时间
mean time to failures（MTTF）   平均无故障时间
mean-square error criterion   均方误差准则
mechanism model   机理模型
meta-knowledge   元知识
minimal realization   最小实现
minimum phase system   最小相位系统
minimum variance estimation   最小方差估计
model confidence   模型置信度
model fidelity   模型逼真度
model reference adaptive control system   模型参考适应控制系统
model verification   模型验证
modern control theory   现代控制理论
modularization   模块化
multi-channel data acquisition   多通道数据采集
multi-channel filtering   多通道滤波
multiloop control   多回路控制
multi-objective decision   多目标决策
multiple-input multiple-output（MIMO）system   多输入多输出系统
multistratum hierarchical control   多段递阶控制
multivariable control system   多变量控制系统
negative feedback   负反馈
neural network   神经网络
neuron   神经元

Nichols chart   尼科尔斯图
nonequilibrium state   非平衡态
nonlinear control system   非线性控制系统
normal mode disturbance   串模干扰
Nyquist stability criterion   奈奎斯特稳定判据
objective function   目标函数
observability index   可观测指数
observable canonical form   可观测规范型
on-line controller   在线控制器
on-off control   通断控制
open loop pole   开环极点
open loop transfer function   开环传递函数
operator station（OS）   操作员站
optical fiber   光缆
optical isolation   光电隔离
optimal control system   最优控制系统
optimal trajectory   最优轨迹
optimization technique   最优化技术
ordinary differential equation   常微分方程
orientation control   定向控制
oscillating period   振荡周期
overall design   总体设计
overdamping   过阻尼
override control   优先控制，超驰控制
overshoot   超调量
parallel negative feedback   并行负反馈
parameter adjustment   参数调节
parameter estimation   参数估计
pattern recognition   模式识别
peak time   峰值时间
perceptron   感知器
performance specification   性能指标
peripheral component interconnection（PCI）bus   外设部件互连总线
phase difference   相位差
phase lag   相位落后
phase lead   相位超前
phase margin   相位裕度
phase-plane method   相平面法

photoelectric isolator　光电隔离器
photoelectric tachometric transducer　光电式转速传感器
pneumatic actuator　气动执行机构
point-to-point control　点位控制
pole assignment　极点配置
pole-zero cancellation　零极点相消
position measuring instrument　位置测量仪
positive feedback　正反馈
precision instrument　精密仪器
pressure compensation control　压力补偿控制
pressure differential meter　差压式流量计
pressure gauge　压力表
pressure transmitter　压力变送器
process automation　过程自动化
process field-bus　过程现场总线
process variable　过程变量
process-oriented simulation　面向过程的仿真
proportional-integral-derivative (PID) controller　比例-积分-微分控制器
pulse transfer function　脉冲传递函数
quadratic performance index　二次型性能指标
qualitative physical model　定性物理模型
ramp function　斜坡函数
random disturbance　随机扰动
ratio control　比值控制
reachability　可达性
reaction control　反作用控制
real display system　实时显示系统
real time control　实时控制
real time database　实时数据库
realizability　可实现性
recorder　记录仪
reduced order observer　降阶观测器
redundancy technique　冗余技术
reference voltage　基准电压
regenerative braking　回馈制动，再生制动
regulating device　调节装置
regulation　调节
relative error　相对误差
relative realiability　相对可靠性
relative stability　相对稳定性
reliability　可靠度
remote manipulator　遥控操作器
resistance thermometer sensor　热电阻
resource allocation　资源分配
response curve　响应曲线
revolution speed transducer　转速传感器
rise time　上升时间
robotics　机器人学
robust control　鲁棒控制
root locus　根轨迹
rotameter　浮子流量计，转子流量计
rotary eccentric plug valve　偏心旋转阀
rotary motion valve　角行程阀
round-off error　舍入误差
Routh-Hurwitz stability criterion　劳斯-霍尔维斯稳定性判据
sampling control system　采样控制系统
sampling holder　采样保持器
saturation characteristic　饱和特性
second-order behavior　二阶特性
selective control　选择控制
self-adaptive control　自适应控制
self-balancing system　自平衡系统
self-organizing system　自组织系统
self-tuning control　自校正控制
semi-physical simulation　半实物仿真
sensitivity analysis　灵敏度分析
sensitivity calibration　灵敏度校准
sensor network　传感器网络
sequential control　顺序控制
serial control unit　串行控制单元
series feedback　串联反馈
servo control　伺服控制
set point value　设定值
settling time　稳定时间

short term planning 短期计划
simulation block diagram 仿真框图
single-input single-output (SISO) system 单输入单输出系统
slower-than-real-time simulation 欠实时仿真
solenoid valve/electromagnetic valve 电磁阀
split-range control 分程控制
stability criterion 稳定性判据
state equation 状态方程
state variable 状态变量
static characteristic 静态特性
steady state deviation 稳态偏差
steady state error coefficient 稳态误差系数
step function 阶跃函数
step response 阶跃响应
step-by-step control 步进控制
stochastic process 随机过程
strongly coupled system 强耦合系统
supervised training 监督学习
supervisory computer control (SCC) 计算机监督控制
switching point 切换点
symbolic processing 符号处理
system assessment 系统评价
system engineering 系统工程
system failure 系统故障
system identification 系统辨识
system simulation 系统仿真
tachometer 转速表
tag name 工位号
target flow transmitter 靶式流量变送器
task cycle 作业周期
temperature regulator 温度调节器
temperature transducer 温度传感器
temperature transmitter 温度变送器
terminal intelligent module 终端智能模块
terminal interface processor 终端接口处理器
thermal sensor 热传感器
thermocouple 热电偶
thermometer 温度计
thickness meter 厚度计
time constant 时间常数
time domain analysis 时域分析
time schedule controller 时序控制器
time-invariant system 定常系统,非时变系统
time-sharing control 分时控制
time-varying parameter 时变参数
tracking error 跟踪误差
trade-off analysis 权衡分析
transducer 传感器
transfer function 传递函数
transient deviation 瞬态偏差
transient response 瞬态响应
transmitter 变送器
trend analysis 趋势分析
turbine flowmeter 涡轮流量计
ultrasonic levelmeter 超声物位计
unbiased estimation 无偏估计
underdamping 欠阻尼
uniformly asymptotic stability 一致渐近稳定性
uninterrupted duty 不间断工作制,长期工作制
unity-feedback system 单位反馈系统
unsupervised learning 非监督学习
variable gain 可变增益,可变放大系数
variable structure control system 变结构控制
velocity error coefficient 速度误差系数
velocity transducer 速度传感器
vibrometer 振动计
voltage source inverter 电压源型逆变器
vortex shedding flowmeter 涡街流量计
weighting factor 权因子
weighting method 加权法
Whittaker-Shannon sampling theorem 惠特克-香农采样定理
work station for computer-aided design 计算机辅助设计工作站

zero-input response	零输入响应	zero	零点
zero-state response	零状态响应	Z-transform	Z 变换

附录 IV　Glossary for Electrical Engineering
（电气工程专业技术词汇）

AC motor　交流电动机
AC transmission system　交流输电系统
AC-DC frequency converter　交-直流变频器
active load　有功负载
active loss　有功损耗
active power　有功功率
adjustable speed motor　调速电动机
aging　老化
alternating current　交流电
ammeter　电流表
anode（cathode）　阳极（阴极）
applied voltage　外加电压
arc discharge　电弧放电
arc reignition　电弧重燃
arc suppression coil　消弧线圈
arc-extinguishing chamber　灭弧室
armature　电枢
asynchronous counter　异步计数器
asynchronous machine　异步电机
asynchronous motor　异步电动机
atomic power generation　原子能发电
attenuation factor　衰减系数
automatic meter reading　自动抄表
automatic voltage regulator（AVR）　自动电压调整器
autotransformer　自耦变压器
bare conductor　裸导线
blackout　断电，停电
boiler　锅炉
breakdown　（电）击穿
brush　电刷
building automation system　楼宇自动化系统
bus bar　母线
bushing　套管
capacitor bank　电容器组
cascade transformer　串级变压器
central control　中央控制室
charging（damping）resistor　充电（阻尼）电阻
circuit breaker　断路器
circuit component　电路元件
circuit parameter　电路参数
clock pulse　时钟脉冲
coaxial cable　同轴电缆
combustion process　燃烧过程
combustion turbine　燃气轮机
commutator　换向器
composite insulation　组合绝缘
compounded　复励的
computer-aided drawing　计算机辅助制图
conductance　电导
conductor　导线
constant current source　恒流源
contactor　接触器
contact　触点
control wiring diagram　控制接线图
copper loss　铜损
core saturation　铁心饱和
corona　电晕
counter　计数器
counter electromotive force　反电动势
coupling capacitor　耦合电容

critical breakdown voltage　临界击穿电压
crossover frequency　交叉频率
current converter　变流器
current ripple　纹波电流
current shunt　分流器
current source inverter　电流（源）型逆变器
current transformer　电流互感器
current-limiting circuit breaker　限流断路器
DC generator-motor set drive　直流发电机-电动机组传动
DC motor　直流电动机
dead tank oil circuit breaker　多油断路器
dead time　死区时间
deenergize　断电，去电
demagnetization　退磁，去磁
demand side management　用电需求管理
demand side management agreement　用电需求管理协议
detection impedance　检测阻抗
dielectric constant　介质常数
dielectric loss　介质损耗
dielectric　电介质，绝缘体
differential protection　差动保护
digital fault recorder　数字故障记录仪
digital voltmeter　数字电压表
discharge　放电
disconnector　隔离开关
dispatch floor　调度楼层
dispatch interval time　调度间隔时间
dispatch panel　调度盘
dispatch signal　调度信号
dispatcher　调度员
dispatcher board　调度程序指示板，综合指示箱
displacement switch　位移开关
distance protection　距离保护
distribution automation system　配电网自动化系统
distribution cabinet　配电柜
distribution circuit　配电线路
distribution dispatch center　配电调度中心
distribution substation　配电站
distribution system　配电系统
distribution transformer　配电变压器
distributor　配电箱
divider ratio　分压器分压比
domestic load　民用电
double bus connection　双母线接线
double-column transformer　双绕组变压器
drive axle　驱动轮轴
drive control　驱动控制
drive motor　驱动电动机
drive system　驱动系统
drive unit　驱动器，驱动单元
dual mode　双机模式
dynamo　直流发电机
earth fault　接地故障
earth (ground) wire　接地线
earthing switch　接地开关
eddy current　涡流
eddy current loss　涡流损耗
effectively grounded　有效接地的
electric field　电场
electrical device　电气设备
Electrical Machinery　电机学
electrochemical deterioration　电化学腐蚀
electromagnetic compatibility　电磁兼容
electromagnetic induction　电磁感应
electromechanical device　机电设备
electronegative gas　电负性气体
electrostatic voltmeter　静电电压表
energy converter　电能转换器
energy management system　能量管理系统
excitation　励磁
excitation effect　励磁效应
excitation system　励磁系统
exciting voltage　励磁电压
exciting winding　励磁绕组

excitor　励磁器
expected life　期望寿命
explosion-proof type　防爆型
expulsion gap　灭弧间隙
extra-high voltage (EHV)　超高压
failure alarming signal　事故报警信号
fault clearing time　故障切除时间
field current　励磁电流
field distortion　场畸变
field strength　场强强度
field stress　电场力
fixed contact　静触头
flash counter　雷电计数器
flashover　闪络
flexible AC transmission system　灵活交流输电系统
flux　磁通
flux density　磁通密度，磁感应强度
fossil-fired power plant　火电厂
full-load torque　满载转矩
gaseous insulation　气体绝缘
generator　发电机
glass insulator　玻璃绝缘子
glow discharge　辉光放电
grounding　接地
grounding capacitance　对地电容
grounding protection　接地保护
harmonic　谐波
heating appliance　电热器
high side　高压侧
high voltage engineering　高电压工程
high voltage testing technology　高电压试验技术
high-voltage winding　高压绕组
hydraulic power generation　水力发电
hydraulic step motor　液压步进马达
hydraulic turbine　水轮机
hydro power station　水力发电站
hydrogenerator　水轮发电机

illuminance　照度
impedance voltage　阻抗电压
impedance　阻抗
impulse current　冲击电流
incoming feeder　进线
induction　感应
induction motor　感应电动机
input winding　输入绕组
instantaneous power　瞬时功率
Institute of Electrical and Electronic Engineers (IEEE)　电气与电子工程师学会（美）
fuse　熔丝
instrument transducer　测量互感器
insulation coordination　绝缘配合
insulator　绝缘子
insulator string　绝缘子串
interlock　互锁设备
intermediate relay　中间继电器
internal combustion engine　内燃机
internal discharge　内部放电
intertrip　联动跳闸
inverter station　换流站
inverter　逆变器
iron core　铁心
iron loss　铁损
kinetic energy　动能
knife switch　刀开关
ladder diagram　梯形图
leakage flux　漏磁通
leakage reactance　漏磁电抗
lighting box　照明箱
lighting main　照明干线
lightning arrester　避雷器
lightning overvoltage　雷电过电压
lightning rod　避雷针
lightning shielding　避雷
lightning stroke　雷电波
line current　线电流
line trap　线路限波器

line voltage　线电压
line-to-neutral　线与中性点间的
live tank oil circuit breaker　少油断路器
load loss　负载损耗
load-saturation curve　负载饱和曲线
local lighting　局部照明
logic diagram　逻辑图
loss angle　（介质）损耗角
lower limit　下限
low-voltage winding　低压绕组
magnetic circuit　磁路
magnetic field　磁场
magnetomotive force　磁通势
main and transfer busbar　单母线带旁路
malfunction　设备故障，失灵
mean free path　平均自由行程
mean molecular velocity　平均分子速度
mechanical accessory　机械部件
megger　兆欧表
milliammeter　毫安表
millivoltmeter　毫伏表
mixed divider　（阻容）混合分压器
motor　电动机
moving contact　动触点
mutual-inductor　互感
negative edge　下降沿
negative ion　负离子
negative sequence impedance　负序阻抗
neutral line　中性线
neutral point　中性点
no-load　空载
no-load current　空载电流
no-load loss　空载损耗
nominal horsepower　额定马力
non-destructive testing　非破坏性试验
non-uniform field　不均匀场
normally closed contact　常闭触点
normally open contact　常开触点
nuclear power station　核电站

oil-impregnated paper　油浸纸绝缘
open circuit voltage　开路电压
operation mechanism　操动机构
operation mode　运行方式
oscilloscope　示波器
output winding　输出绕组
over current protection　过电流保护
overflux　过励磁
overhead line　架空线
overvoltage　过电压
partial discharge　局部放电
peak reverse voltage　反向峰值电压
peak voltmeter　峰值电压表
peak-load　峰荷
performance test　性能测试
periodic duty service　定期运行
permanent magnet　永磁体
phase current　相电流
phase displacement（shift）　相移
phase shifter　移相器
phase voltage　相电压
phase-to-phase voltage　线电压
photoelectric emission　光电发射
plug-in type fuse　插入式熔断器
polarity effect　极性效应
polyphase rectifier　多相整流器
porcelain insulator　陶瓷绝缘子
positive edge　上升沿
positive sequence impedance　正序阻抗
potential stress　电位应力（电场强度）
potential transformer（PT）　电压互感器
potentionmeter　电位器
power balance　功率平衡
power capacitor　电力电容
power connector　电源接头
power electronics　电力电子
power factor　功率因数
power factor meter　功率因数表
power line carrier（PLC）　电力线载波（器）

power network 电力网络
power plant 电厂
power supply box 动力箱
power supply line 供电线路
power system 电力系统
power system automation 电力系统自动化
power transformer 电力变压器
power transmission system 输电系统
primary grid 主网
primary/secondary mode 主/从模式
primary（backup）relaying 主（后备）继电保护
primary winding 一次绕组
prime grid substation 主网变电站
programmable logic controller（PLC） 可编程序逻辑控制器
pull switch 拉线开关
pulse width modulation control system 脉冲调宽控制系统
PWM inverter 脉宽调制逆变器
radio frequency 射频
radio interference 无线干扰
rated capacity 额定容量
rated current 额定电流
rated load 额定负载
rated power 额定功率
rated speed 额定转速
rated voltage 额定电压
rating of equipment 设备额定值
reactance 电抗
reactive 电抗的，无功的
reactive load 无功负载
reactive loss 无功损耗
reactive power 无功功率
reactive power compensation 无功补偿
reactive power optimization 无功优化
reactor 电抗器
reclosing 重合闸
recovery voltage 恢复电压

rectifier 整流器
redundancy mode 冗余模式
regenerative braking 回馈制动，再生制动
relay characteristic 继电器特性
relay panel 继电器屏
relay protection 继电保护
relay 继电器
reluctance 磁阻
remote terminal unit （RTU）远程终端设备
reserve capacity 备用容量
residual capacitance 残余电容
rheostat 变阻器
right-of-way 线路走廊
root mean square 方均根值
rotating magnetic field 旋转磁场
rotor 转子
routine test 常规试验
routing testing 常规试验
rum 炉筒
saturation curve 饱和曲线
saturation effect 饱和效应
saturation factor 饱和系数
secondary substation 二次变电站
secondary winding 二次绕组
self-diagnostics 自诊断
self-inductance 自感
separately excited 他励的
sequential tripping 顺序跳闸
series（shunt）compensation 串（并）联补偿
series capacitor compensation 串联电容补偿
series motor 串励电动机
series winding 串联绕组
servo control 伺服控制，随动控制
servomechanism 伺服系统
servomotor 伺服电动机
shielding 屏蔽
short circuit testing 短路试验
short-circuit ratio 短路比

short-circuiting ring 短路环
shunt capacitor 并联电容器
shunt field 并励磁场
shunt motor 并励电动机
shunt reactor 并联电抗器
signal indicator 信号指示器
silicon-controlled rectifier 晶闸管整流器
single-phase asynchronous motor 单相异步电动机
skin effect 趋肤效应
slip ring 集电环
solar energy power generation 太阳能发电
space charge 空间电荷
speed control system 调速系统
starter 起动器
starting current 起动电流
starting torque 起动转矩
stationary blade 固定叶片
stator 定子
steady state 稳态
steady-state analysis of power system 电力系统稳态分析
steam turbine 汽轮机
steel-reinforced aluminum 钢芯铝绞线
step down transformer 降压变压器
step up transformer 升压变压器
storage battery 蓄电池
stray capacitance 杂散电容
streamer breakdown 流注击穿
substation 变电站
subtracting counter 减法计数器
superheater 过热器
supervisory control and data acquisition (SCADA) 监控与数据采集
surface breakdown 表面击穿
surge 冲击，过电压
surge impedance 波阻抗
susceptance 电纳
suspension insulator 悬式绝缘子
sustained discharge 自持放电
switchboard 配电盘，开关屏
switching overvoltage 操作过电压
symmetrical three-phase load 对称三相负载
symmetrical three-phase source 对称三相电源
synchronous condenser 同步调相机
synchronous counter 同步计数器
synchronous motor 同步电动机
synchronous reactance 同步电抗
synchronous speed 同步转速
tap 分接头
tap position 档位
taped transformer 多级变压器
technical specification 技术规格
terminal block 端子排
thermal breakdown 热击穿
thermal overload relay 热继电器
thermal power station 火力发电站
thermal rating 额定温度
three-phase fault 三相故障
three-column transformer 三绕组变压器
three-phase asynchronous motor 三相异步电动机
three-phase circuit 三相电路
three-phase four-wire system 三相四线制
three-phase power 三相功率
three-phase source 三相电源
three-phase three-wire system 三相三线制
tidal current 潮流
tidal power generation 潮汐发电
time relay 时间继电器
time-lag element 延时元件
timing diagram 时序图
timing simulation 时序模拟
torque 力矩
transformer 变压器
transformer substation 变电站
transient state 暂态
transient-state analysis of power system 电力

系统暂态分析
transmission line　输电线
travel switch　行程开关
trial operation　试运行
triangular connection　三角形联结
trigger electrode　触发电极
trigger pulse　触发脉冲
trip circuit　跳闸电路
trip coil　跳闸线圈
tuned circuit　调谐电路
turbogenerator　汽轮发电机
turn　匝
turn ratio　匝比，变比
ultra-high voltage (UHV)　特高压
underground cable　地下电缆
uniform field　均匀场
uninterruptible power supply　不间断电源
upper limit　上限
utilization equipment　用电设备

vacuum contactor　真空接触器
vacuum circuit breaker　真空断路器
variable transformer　调压变压器
voltage divider　分压器
voltage drop　电压降
voltage grade　电压等级
voltage regulation　电压调节
voltage stability　电压稳定性
watt-hour meter　电能表
wave distortion　波形失真
wave front　波头
wave tail　波尾
winding　绕组
withstand voltage　耐受电压
work station　工作站
zero potential　零电位
zero sequence current　零序电流
zero sequence impedance　零序阻抗

附录 V　Glossary for Computer Science
（计算机专业技术词汇）

access　存取
access time　存取时间
active statement　活动语句
adapter cards　适配卡
add-in　加载项
add-ons　插件
advanced application　高级应用
aggregate query　聚合查询
animation　动画
application programming interface (API)　应用程序接口
application software　应用软件
archive file　存档文件
arithmetic operation　算术运算

arithmetic-logic unit (ALU)　算术逻辑单元
asynchronous communications port　异步通信端口
attachment　附件
audio-output device　音频输出设备
authentication　身份验证
authorization　授权
automatic recovery　自动恢复
back up　备份
backup device　备份设备
backup file　备份文件
balanced hierarchy　均衡层次结构
bar code　条形码
bar code reader　条形码读卡器

basic application　基础程序
batch　批处理
binary coding scheme　二进制译码方案
binary data type　二进制数据类型
binary system　二进制系统
binding　绑定
bitwise operation　按位运算
bit　比特，位
bluetooth　蓝牙
broadband　宽带
browser　浏览器
business-to-business　企业对企业
cable modem　有线调制解调器
central processing unit（CPU）　中央处理器
chain printer　链式打印机
channel　通道
character and recognition device　字符标志识别设备
chat group　谈话群组
closed architecture　封闭式体系结构
commerce server　商业服务器
communication channel　信道
compact disc（CD）　压缩盘
computer crime　计算机犯罪
computer network　计算机网络
computer support specialist　计算机支持专家
computer technician　计算机技术人员
connection device　连接设备
continuous-speech recognition system　连续语言识别系统
control unit　操纵单元
control-break report　控制中断报表
control-of-flow language　控制流语言
cookies-cutter program　信息记录截取程序
cookies　信息记录程序
cordless or wireless mouse　无线鼠标
cracker　解密高手
custom rule　自定义规则
cyberspace　计算机空间

data block　数据块
data bus　数据总线
data connection　数据连接
data definition language（DDL）　数据定义语言
data explosion　数据爆炸
data integrity　数据完整性
data manipulation language（DML）　数据操作语言
data modification　数据修改
data projector　数码放映机
data pump　数据抽取
data scrubbing　数据清理
data security　数据安全
data source name（DSN）　数据源名称
data transmission specification　数据传输说明
data warehouse　数据仓库
database administrator　数据库管理员
database catalog　数据库目录
database diagram　数据关系图
database management system（DBMS）　数据库管理系统
database schema　数据库架构
database script　数据库脚本
data-definition query　数据定义查询
dataplay　数字播放器
dataset　数据集
decimal data type　十进制数据类型
decision support　决策支持
decision tree　决策树
default database　默认数据库
default instance　默认实例
delete query　删除查询
desktop system unit　台式计算机系统单元
destination file　目标文件
dial-up service　拨号服务
differential database backup　差异数据库备份
digital camera　数码照相机
digital cash　数字现金

digital notebook 数字笔记本
digital subscriber line 数字用户线路
digital video camera 数码摄影机
digital video disc 数字化视频光盘
dimension hierarchy 维度层次结构
direct access 直接存取
direct response mode 直接响应模式
directory search 目录搜索
discrete-speech recognition system 不连续语言识别系统
disk caching 磁盘驱动器高速缓存
distributed data processing system 分布数据处理系统
distributed processing 分布处理
distributed query 分布式查询
distributor 分发服务器
document file 文档文件
domain code 域代码
domain integrity 域完整性
domain name system（DNS） 域名服务器
dot-matrix printer 点矩阵式打印机
double-byte character set（DBCS） 双字节字符集
dual-scan monitor 双向扫描显示器
dumb terminal 非智能终端
dump file 转储文件
dynamic cursor 动态游标
dynamic filter 动态筛选
dynamic locking 动态锁定
dynamic recovery 动态恢复
dynamic snapshot 动态快照
e-book 电子阅读器
electronic cash 电子现金
electronic commerce 电子商务
encrypted trigger 加密触发器
encryption 加密
end user 终端用户
enterprise computing 企业计算化
entity integrity 实体完整性
erasable optical disk 可擦除式光盘
error log 错误日志
error state number 错误状态号
escape character 转义符
expansion card 扩展卡
extended stored procedure 扩展存储过程
external modem 外置调制解调器
extranet 企业外部网
fax machine 传真机
fetch 提取
fiber-optic cable 光纤电缆
field 域
field length 字段长度
field terminator 字段终止符
file compression 文件压缩
file decompression 文件解压缩
file storage type 文件存储类型
file transfer protocol（FTP） 文件传送协议
filtering 筛选
firewall 防火墙
flash memory 闪存
flat-panel monitor 纯平显示器
flexible disk 可折叠磁盘
float data type 浮点数数据类型
floppy disk 软盘
formatting toolbar 格式化工具条
forward-only cursor 只进游标
full-duplex communication 全双通通信
full-text catalog 全文目录
full-text index 全文索引
full-text query 全文查询
general-purpose application 通用程序
global default 全局默认值
global property 全局属性
global variable 全局变量
graphic tablet 绘图板
graphical user interface 图形用户界面
hacker 黑客
handheld computer 手提电脑

hard copy   硬拷贝
hard disk   硬盘
hard-disk pack   硬盘组
hardware   硬件
header   标题
help desk specialist   帮助办公专家
helper applications   帮助软件
heterogeneous data   异类数据
hierarchical network   层次型网络
high performance serial bus（HPSB）   高性能串行总线
high-definition television（HDTV）   高清电视
home page   主页
homogeneous data   同类数据
host computer   主机
hot site   热点网站
hybrid network   混合网络
hyperlinks   超连接
hypertext markup language（HTML）   超文本链接标志语言
index search   索引搜索
identity property   标志属性
idle time   空闲时间
image capturing device   图像获取设备
image data type   图像数据类型
immediate updating subscriber   即时更新订阅服务器
immediate updating   即时更新
implied permission   暗示性权限
incremental update   增量更新
index page   索引页
information model   信息模型
information technology   信息技术
initializing   初始化
ink-jet printer   墨水喷射印刷机
inner join   内联接
insensitive cursor   不感知游标
Insert query   插入查询
instant messaging   即时信息

integer data type   整数数据类型
integrated package   综合性组件
integrated security   集成安全性
integrity constraint   完整性约束
intelligent terminal   智能终端设备
interface card   接口卡
internal hard disk   内置硬盘
internal identifier   内部标志符
internal modem   内部调制解调器
Internet hard drive   网络硬盘驱动器
Internet relay chat（IRC）   互联网多线交谈
Internet telephony   网络电话
Internet terminal   互联网终端
intranet   企业内部网
join condition   连接条件
join operator   连接运算符
join path   连接路径
join table   连接表
joystick   操纵杆
key range lock   键范围锁
keyword search   关键字搜索
laser printer   激光打印机
level hierarchy   级别层次结构
light pen   光笔
linked cube   链接多维数据集
linked server   链接服务器
linking table   链接表
liquid crystal display monitor（LCD）   液晶显示器
local area network（LAN）   局域网
local cube   本地多维数据集
local distributor   本地分发服务器
local group   本地组
local login identification   本地登录标志
local server   本地服务器
local subscription   本地订阅
local variable   局部变量
locale identifier（LCID）   区域设置标志符
log file   日志文件

logical name　逻辑名称
logical operation　逻辑运算
logical operator　逻辑运算符
login security mode　登录安全模式
magnetic tape reels　磁带卷
magnetic tape streamer　磁带条
main board　主板
make table query　生成表查询
many-to-many relationship　多对多关系
many-to-one relationship　多对一关系
mark sensing　标志检测
mechanical mouse　机械鼠标
member group　成员组
member property　成员属性
memory　内存
merge replication　合并复制
message number　消息编号
messaging application programming interface（MAPI）　消息应用程序接口
meta data　元数据
metropolitan area network（MAN）　城域网
microprocessor　微处理器
microseconds　微秒
mining model　挖掘模型
mirroring　镜像
mixed mode　混合模式
model dependency　模型相关性
modem　解调器
monitor　显示器
motherboard　主板
multiple inheritance　多重继承
multithreaded server application　多线程服务器应用程序
multiuser　多用户
national service provider　全国性服务供应商
native format　本机格式
nested query　嵌套查询
nested table　嵌套表
network adapter card　网卡

network architecture　网络体系结构
network bridge　网桥
network gateway　网关
network manager　网络管理员
network terminal　网络终端
newsgroup　新闻组
notebook computer　笔记本式计算机
notebook system unit　笔记本系统单元
numeric entry　数字输入
numeric expression　数值表达式
object dependencies　对象相关性
object embedding　对象嵌入
object identifier　对象标志符
object linking and embedding（OLE）　对象链接嵌入
object permission　对象权限
off-line browser　离线浏览器
one-to-many relationship　一对多关系
one-to-one relationship　一对一关系
online analytical processing（OLAP）　联机分析处理
online redo log　联机重做日志
online storage　联机存储
online transaction processing（OLTP）　联机事务处理
open architecture　开放式体系结构
open data services（ODS）　开放式数据服务
open database connectivity（ODBC）　开放式数据库连接
open information model（OIM）　开放信息模型
operation system　操作系统
optical mouse　光电鼠标
optical scanner　光电扫描仪
optical-character recognition（OCR）　光电字符识别器
optical-mark recognition（OMR）　光标阅读器
optimize synchronization　优化同步
outer join　外连接
packet　数据包

page split 页拆分
palmtop computer 掌上电脑
parallel data transmission 平行数据传输
parallel port 并行端口
peer-to-peer network system 点对点网络系统
peripheral component interconnect（PCI） 外部设备互连总线
personal digital assistant（PDA） 个人数字助理
personal laser printer 个人激光打印机
personal video recorder card 个人视频记录卡
physical name 物理名称
platform scanner 平版式扫描仪
plotter 绘图仪
plug and play 即插即用
plug-in board 插件卡
portable scanner 便携式扫描仪
positioned update 定位更新
presentation file 演示文稿
presentation graphic 电子文稿程序
primary storage 主存
programming control language 程序控制语言
property page 属性页
proxy 代理服务器
proxy server 代理服务
publication database 发布数据库
publication retention period 发布保持期
published data 已发布数据
publisher 发布服务器
pull subscription 请求订阅
push subscription 强制订阅
query optimizer 查询优化器
question builder 问题生成器
random access memory（RAM） 随机存储器
read-only memory（ROM） 只读存储器
real data type 实数数据类型
refresh data 刷新数据
regional service provider 区域性服务供应商
relational database 关系数据库
relational database management system（RDBMS） 关系数据库管理系统
relationship object 关系对象
remote distributor 远程分发服务器
remote login identification 远程登录标志
remote partition 远程分区
remote server 远程服务器
remote stored procedure 远程存储过程
replication conflict viewer 复制冲突查看器
replication monitor 复制监视器
repository 知识库
repository engine 知识库引擎
resolution strategy 冲突解决策略
resolution 分辨率
reusable bookmark 可再次使用的书签
right outer join 右向外连接
ring network 环形网
roll back 回滚
roll forward 前滚
row filter 行筛选
row lock 行锁
sample data 示例数据
scalar aggregate 标量聚合
scanner 扫描器
scheduled backup 已调度备份
schema 架构
standard security 标准安全机制
standard toolbar 标准工具栏
star join 星形联结
star network 星形网
star schema 星形架构
statement permission 语句权限
static cursor 静态游标
store-and-forward database 保存与转发数据库
string 字符串
string functions 字符串函数
structured query language（SQL） 结构化查询语言
structured storage file 结构化存储文件

subquery　子查询
subscriber　订阅服务器
subscription address　预定地址
supercomputer　巨型计算机
superlative form　最高级
surfing　网上冲浪
surge protector　浪涌保护器
synchronization　同步
system administrator　系统管理员
system catalog　系统目录
system clock　时钟
system software　系统软件
system stored procedure　系统存储过程
systems analyst　系统分析师
table lock　表锁
table scan　表扫描
tabular data stream（TDS）　表格格式数据流
tape backup　磁带备份
target partition　目标分区
task object　任务对象
television board　电视扩展卡
telnet　远程登录
template　模板
temporary stored procedure　临时存储过程
text data type　文本数据类型
text entry　文本输入
thin film transistor monitor（TFT）　薄膜晶体管显示器
time-sharing system　分时系统
toggle key　触发键
topological structure　拓扑结构
touch screen　触摸屏
trace file　跟踪文件
traditional cookies　传统的信息记录程序
transaction log　事务日志
transaction processing　事务处理

transaction rollback　事务回滚
transactional replication　事务复制
transformable subscription　可转换订阅
trusted connection　信任连接
twisted pair　双绞线
universal serial bus（USB）　通用串行总线
video display screen　视频显示屏
virtual memory　虚拟内存
virus checker　病毒检测程序
voltage surge　电涌
web appliance　环球网设备
web broadcasters　网络广播
web page　网页
web portals　门户网站
web server　服务器
web site　网站
web site address　网络地址
web storefronts　网上商店
web terminal　环球网终端
web utilities　网上应用程序
webcam　摄像头
web-downloading utilities　网页下载应用程序
webmaster　站点管理员
wide area network（WAN）　广域网
wildcard character　通配符
window menu　窗口菜单
wireless modem　无线调制解调器
wireless service provider　无线服务供应商
word processing　文字处理
word wrap　自动换行
World Wide Web　万维网
worm　蠕虫病毒
write-protect notch　写保护口
zip disk　压缩磁盘
zone list　区域列表
zone transfer　区域传送

附录 VI Expression of Numerals, Mathematic Symbols and Formulas
（数字、符号和公式的表达）

1. 小数和百分数的读法

小数点读为 point [pɔɪnt]，小数点前的整数读基数词，也可将每个数字按位分开读。一般读法是：二位数以内可合读一个基数词，三位数以上的，多按每个数分开读；小数点后的小数部分，均采取每个数字分开读的方法读。例如：

0.8 读为 point eight 或 o [ou] point eight。

2.32 读为 two point three two。

40.51 读为 fourty point five one。

145.678 读为 one four five point six seven eight。

读英语百分数时，只要在基数词后加上 percent 就行。例如：

45% 读为 forty-five percent。

0.2% 读为 point two percent。

5.33% 读为 five point three three percent。

2. 分数的读法

英语分数是以基数词和序数词合成的。基数词代表分子，序数词代表分母。除分子是 1 的分数外，序数词一般都要读为复数。例如：

$\frac{1}{5}$ 读为 one fifth。

$1\frac{1}{4}$ 读为 one and one fourth 或 one and a quarter。

$2\frac{7}{12}$ 读为 two and seven twelfths。

分数往往有按比的读法，如 $\frac{2}{3}$ 或 2 比 3 读为 two over three。此外，$\frac{1}{2}$ 和 $\frac{1}{4}$ 通常可读为 one (a) half 和 one (a) quarter。

3. 年、月、日的读法

科技英语中出现的年、月、日的读法与普通英语一样，其年份的读法有两种：

In 1898 读为 in eighteen ninety-eight。

In 1980 读为 in nineteen eighty。

In 1900 读为 in nineteen hundred 或读为 in nineteen double 0 [əʊ]。

In 2017 读为 in two thousand and seventeen 或者 twenty seventeen。

月份按音标读出其名称即可；日期一般读序数词，而且前面要加定冠词 the。例如：

五月一日读为 May the First 或 the First of May。

十二月七日读为 December the seventh。

年月日在一起表示一个具体日子时，在最前面一般用介词 on，顺序是"月份 + 日期 + 年份"。例如：

on October 1, 1979 读为 on October (the) first, nineteen seventy-nine。

on May 4, 1981 读为 on May (the) fourth, nineteen eight-one。

4. 书刊的章节及各种数码的读法

英语书刊的章节一般也有两种读法：一种是"名词 + 基数词"；另一种是"定冠词the + 序数词 + 名词"。例如：

第二节读为 section two 或读为 the second section。

第四章读为 chapter four 或读为 the fourth chapter。

第十八课读为 lesson eighteen 或读为 the eighteen 或读为 the eighteenth lesson。

Type 301 读为 type three 0 [əʊ] one。

Room 212 读为 room two one two。

3-7085 读为 three seven 0 [əʊ] eight five。

5. "0" 的读法

数字"0"（零）这个符号有以下几种读法：

1）"0"在数字中如要单个读出时，可以读为 [əʊ] 或 zero 或 nought [nɔːt]。

506.02 读为 five o (əʊ) six point o (əʊ) two。

但如果在整数中，一般只读基数词的名称，不读出零。

10 读为 ten。

106 读为 one hundred and six。

1005 读为 one thousand and five。

2）"0"作为数值读出时，还可以读为 cipher 或 cypher [ˈsaɪfə] 和 nothing。例如：

10 − 10 = 0 读为 Ten minus ten is equal to cipher。或读为 Ten minus ten leaves nothing。

5 × 0 = 0 读为 Multiply 5 by nothing and the result is nothing。

3）"0"在表示温度的"零上"或"零下"时，一般应读为 zero。例如：

0℃ 读为 zero Centigrade。

−3℃ 读为 three degrees below zero Centigrade，也可读为 minus three degrees。

4）"0"在号码及年份中，一般都读为 [əʊ]。例如：：Room 205 读为 Room two 0 [əʊ] five。

Dial 20050 读为 Dial two double 0 [əʊ] five 0 [əʊ]。

In 1907 读为 in nineteen 0 [əʊ] seven。

In 1900 读为 in nineteen double 0 [əʊ]。

5）"0"在运动会中的比分一般读为 nil [nɪl]，或 nothing。例如：

3∶0 读为 three∶nil 或 three∶nothing；也可读为 three goals to nil。

5∶0 读为 five∶nothing。

6. 常用数学符号的读法

≌ 读为 be congruent with 或 approximately equal。

~ 读为 equivalent to。

π 读为 pi or the ratio of the circumference of a circle to the diameter。

∞ 读为 infinity。
∝ 读为 varies directly as。
$n!$ 读为 the factorial of n。
e = 2.7182818285... 读为 the base of natural logarithm。
lg a 读为 lg common logarithm。
ln a 读为 natural logarithm。
Δy 读为 the increment of y。
dy 读为 the differential of y。
$\sum_{i=1}^{n}$，$\sum_{1}^{n}$ 读为 sum to n terms。
$\prod_{i=1}^{n}$ 读为 product of a teems。
$F(x)$ 读为 the function of x。
lim sup 读为 superior limit。
lim inf 读为 inferior limit。
$\frac{dy}{dx}$ 读为 the derivative of y with respect to x。
$\frac{d^n y}{dx^n}$ 读为 the nth derivative of y with respect to x。
$\frac{\partial f}{\partial x}$ 读为 the partial derivative of f with respect to x。
$\int f(x)dx$ 读为 the integral of $f(x)$ with respect to x, the primitive of $f(x)$。
$\int_a^b f(x)dx$ 读为 the definite integral of $f(x)$ from a to b。
$I(z)$，$\text{Im}(z)$ 读为 the imaginary part of z。
$R(z)$，$\text{Re}(z)$ 读为 the real part of z。
$\overline{Z}$ 读为 the conjugate of z (z 的共轭复数)。
> 读为 greater than。
< 读为 less than。
= 读为 equals/is equal to。
/ 读为：over。
$\sqrt{x}$ 读为 the square/second root of x。
$\sqrt[3]{x}$ 读为 the cube/third root of x。
x^2 读为 x squared 或者 the second power of x。
x^3 读为 x cubed 或者 the third power of x。

7. 公式的读法

3 + 4 = 7 读为 Three plus four equals (is equal to) seven。
8 - 5 = 3 读为 Eight minus five equals (is equal to) three。
2 × 3 = 6 读为 Two times three equals six; Two times three is equal to six; Two multiply by

three is six。

$9 \div 3 = 3$ 读为 Nine divided by three equals (is equal to) three。

$2^2 = 4$ 读为 two squared equals four。

$4^3 = 64$ 读为 four cubed equals sixty-four。

3^{12} 读为 three to the twelfth power。

10^{-5} 读为 ten to the minus five。

$j = \sqrt{-1}$ 读为 J (dʒei) equals the square root of minus one。

$10 \times 8\text{feet}$ 读为 ten by eight feet。

$y : \dfrac{1}{x^2}$ 读为 y is proportion to one over x squared；或读为 y varies inversely as x squared。

$a^2 = b^2 + c^2$ 读为 a squared equals b squared plus c squared。

$i = \sqrt{-1}$ 读为 the unit of imaginary numbers。

$x = \dfrac{a}{y} + \dfrac{b}{y}$ 读为 x equals a over y plus b over y。

$y = \pm\sqrt{b - x^2}$ 读为 y equals plus-or-minus the square root of b minus x squared。

$S_1 = S_2 - ak^2$ 读为 S sub one equals S sub two minus ak squared。

$y^2 \geq 4$ 读为 y squared is equal to or greater than four。

$y \leq 1 - \dfrac{1}{2}x^2$ 读为 y is equal to or less than one minus a half x squared。

$P = \dfrac{K}{V}$ 读为 P equals K divided by V 或 P is equal to K over V。

$F = ma$ 读为 F equals m times a；或 F equals ma。

$I = \dfrac{U}{R}$ 读为 I equals U over R 或 U divided by R。

$a \ll b$ 读为 a is much less than b。

$a \gg b$ 读为 a is much/far greater than b。

参 考 文 献

[1] 魏汝尧,董益坤.致用科技英语[M].北京:国防工业出版社,2007.
[2] 秦荻辉.科技英语(电子类)[M].西安:西安电子科技大学出版社,2008.
[3] 谢屏,桂清扬.21世纪科技英语:下册[M].北京:高等教育出版社,2002.
[4] 戴文进.科技英语翻译理论与技巧[M].上海:上海外语教育出版社,2007.
[5] 杨跃,马刚.实用科技英语翻译研究[M].西安:西安交通大学出版社,2008.
[6] 闫文培.实用科技英语翻译要义[M].北京:科学出版社,2008.
[7] Paulette Dale, James C Wolf. Speech Communication Made Simple [M]. 3rd ed. London: Person Education, Inc., 2006.
[8] 严俊仁.汉英科技翻译新说[M].北京:国防工业出版社,2010.
[9] 秦荻辉.科技英语写作教程[M].西安:西安电子科技大学出版社,2001.
[10] 严俊仁.英汉科技翻译新说[M].北京:国防工业出版社,2010.
[11] 严俊仁.900科技英语长难句分析与翻译[M].北京:国防工业出版社,2010.
[12] Alan V Oppenheim, Alan S Willsky, S Hamid Nawab. Signals and Systems. [M]. 2nd ed. 北京:电子工业出版社,2005.
[13] 任胜利.英语科技论文撰写与投稿[M].北京:科学出版社,2004.
[14] 童玉珍.牛津电子学英语[M].北京:北京大学出版社,1998.
[15] Basil Hamed. Design & Implementation of Smart House Control Using LabVIEW [J]. International Journal of Soft Computing and Engineering, 2012, 1 (6): 98-106.
[16] O Adeogun, A Tiwari, J R Alcock. Models of Information Exchange for UK Telehealth Systems [J]. International Journal of Medical Informatics, 2011 (80): 359-370.
[17] Rossi D T, Zhang N. Automating Solid-Phase Extraction: Current Aspects and Future Prospects [J]. Journal of Chromatography, 2000, 885 (1-2): 97-113.
[18] Ghassan Hamarneh, Judith Hradsky. Bilateral Filtering of Diffusion Tensor Magnetic Resonance Images [J]. IEEE Transactions on Image Processing, 2007, 16 (10): 2463-2475.
[19] Waikai Chen. Circuit Analysis and Feedback Amplifier Theory [M]. Florida: CRC press, 2005.
[20] Steven W Smith. The Scientist and Engineer's Guide to Digital Signal Processing [M]. California Technical Publishing, 1997.
[21] Pavel Hamet, Johanne Tremblay. Artificial Intelligence in Medicine [J]. Metabolism Clinical and Experimental, 2017, (69): 36-40.
[22] Mehmed Kantardzic. Data Mining: Concepts, Models, Methods, and Algorithms [M]. 2nd ed. New York: John Wiley & Sons, Inc., 2011.
[23] Mohammad Derawi, Hao Zhang. Internet of Things in Real-Life—A Great Understanding. Q.-A. Zeng (ed.), Wireless Communications, Networking and Applications, Lecture Notes in Electrical Engineering [M]. New Delhi: Springer India, 2016.
[24] Serbedziji N. Adaptive Assistance: Smart Home Nursing. In: Nikita K S. Lin J C, Fotiadis D I, Arredondo Waldmeyer MT. (eds) Lecture Notes of the Institute for Computer Sciences, Social Informatics and Telecommunications Engineering [C]. Berlin: Springer-Verlag, 2012: 240-247.
[25] Sarra Hammoudi, Zibouda Aliouat, Saad Harous. Challenges and Research Directions for Internet of Things [J]. Telecommunication Systems, 2017: 1-19.

［26］Haibin Yan, Marcelo H Ang Jr., Aun Neow Poo. A Survey on Perception Methods for Human-Robot Interaction in Social Robots［J］. Int. J. Soc. Robot, 2014（6）：85-119.

［27］Sue Ellen Haupt, Antonello Pasini, Caren Marzban. Artificial Intelligence Methods in the Environmental Sciences［M］. Berlin：Springer-Verlag, 2009.

［28］Damian Flynn. Thermal Power Plant Simulation and Control［M］. London：The Institution of Engineering and Technology, 2003.

［29］Richard Crowder. Electric Drives and Electromechanical Systems：Applications and Control［M］. Boston：Newnes, 2006.

［30］Shankar P Bhattacharyya, Aniruddha Datta, L H Keel. Linear Control Theory［M］. Florida：CRC Press, 2009.

［31］Fernando D Bianchi, Hernan De Battista, Ricardo J Mantz. Wind Turbine Control Systems［M］. Berlin：Springer-Verlag, 2003.

［32］Francis Bacon Crocker, Morton Arendt. Electric Motors, Their Action, Control and Application［M］. Philadelphia：Nabu Press, 2010.

［33］Benjiamin G Kuo, Farid Golnaraghi. Automatic Control Systems［M］. 8th ed. New York：Jonhn Wiley Inc., 2001.

［34］徐勇, 王志刚. 英语科技论文翻译与写作教程［M］. 北京：化学工业出版社, 2015.

［35］赵志毅. 科技英语三百问［M］. 西安：陕西人民出版社, 1985.

［36］吴建, 张韵菲. 汉英科技翻译教程［M］. 南京：南京大学出版社, 2014.

［37］任胜利. 英语科技论文撰写与投稿［M］. 北京：科学出版社, 2015.

［38］魏羽, 高宝萍. 汉英科技翻译教程［M］. 西安：西北工业大学出版社, 2010.